To Paul Kennedy,
with thanks,
Ken Moynihan
Jan. 22, 2004

Assumption College

A Centennial History

1904-2004

Kenneth J. Moynihan

ASSUMPTION COLLEGE
CENTENNIAL 2004

Published by Assumption College, 2004

Printed by Thomson-Shore, Inc., Dexter, Michigan

Published by Assumption College, 2004
500 Salisbury Street, Worcester, MA 01609

Printed in the United States of America

ISBN 0-9654033-0-0

To Wilfrid J. Dufault, A.A.

and to the memory of Raymond J. Marion,

representing the best of both worlds.

Acknowledgements

I am grateful to President Joseph H. Hagan for selecting me to prepare this history, for providing me with the time to carry out the task free of other duties, and for granting me full editorial control of the text. President Thomas R. Plough has honored those agreements and provided personal support. Very Rev. John L. Franck, A.A. made much of the project possible by granting me access to the archives of the Assumptionists' North American Province, housed in Brighton, Massachusetts. There I received unfailing help and encouragement from Fr. Richard Richards, A.A. as he shared his collections and his profound knowledge of Assumptionist history with me.

Very Rev. Wilfrid J. Dufault, A.A., by voluntarily vouching for me, helped me gain access to the rich archives of the Assumptionist General House in Rome. For that access I am also grateful to Very Rev. Richard E. Lamoureux, A.A., superior general, and to Fr. Jean-Paul Périer-Muzet, A.A., archivist. Fr. Noël Bugnard, archivist of the Paris Province, scoured his collections for documents left behind when the records regarding North America were transferred to the United States decades ago. He found and sent a small but significant collection. At Assumption College, Fr. Donat Lamothe, A.A., college archivist, provided me with full access, the key to the door, and frequent words of welcome advice. For many kindnesses, I am grateful to Dawn R. Thistle, director of library services, and the staff of the Emmanuel d'Alzon Library.

Leslie P. Choquette, Charles W. Estus, Sr. and Mary Jo Moynihan read the chapters as they emerged and offered suggestions that significantly improved the text. Caroline F. Sloat's expert editing and enthusiasm for the project were a great help. The completed text was prepared for publication by a quartet of women to whom I am deeply indebted: Sheila McAvey did the proofreading and created the index; Kim Cutler designed the book, set the type, and scanned in the pictures; Victoria Hughes Waters gave the text a final proofreading; Nancy McBride resourcefully played the many roles required of a publisher.

Although two other names appear on the dedication page, this work is likewise dedicated to my wife, Mary Jo Moynihan, to whom my debt is beyond words.

Table of Contents

Assumption College

A Centennial History

1904–2004

Kenneth J. Moynihan

Prologue

The Road to Worcester (1810–1903)

The story of Assumption College began when members of the Augustinians of the Assumption, a Roman Catholic religious congregation founded in France, came into contact with a population that shared both their Catholic faith and their French language. The fruitful connection occurred at an unlikely spot, a Protestant, English-speaking city in New England. Assumptionist and French-Canadian pathways converged in Worcester, Massachusetts, in 1903. For decades before that, the foundations of their unique relationship were being laid on both sides of the Atlantic.

Worcester in 1870 was a burgeoning industrial city distinguished principally by the diversity of the products flowing from its shops and factories. Although more than one-fifth of the work force was employed by the declining boot and shoe industry, two manufacturing groups, "metals and metallic goods" and "machines and machinery" were reasserting their earlier prominence in the industrial composition of the community. The manufacture of textiles — also a declining sector of the economy by 1870 — continued to occupy large numbers of workers. Many were also employed in the making of envelopes, leather belting, carpets, and abrasives.[1]

Notre Dame des Canadiens Church in 1870 (Insert: St. Anne's Church, begun by Canadians in 1852; lost due to financial failure)

In the 1870 United States census, the city was officially credited with a population of 41,105, an increase of 65 percent in the course of a decade. The residents were mostly Protestant in religion, descended from English and Scottish forebears. French-speaking citizens of Canada had been flowing into Worcester over the two decades prior to 1870, adding to an influx of Irish-Catholic immigrants that had begun in the 1820s with workers on the Blackstone Canal. By 1869, the French Canadians, under the leadership of Fr. Jean-Baptiste Primeau, had organized their first permanent parish, Notre Dame des Canadiens, which met in rented space in Horticultural Hall, across from City Hall on Front Street. In 1870, it moved across the Common to take possession of what had been a Methodist church on Park Street (later Franklin Street). According to its own census, in 1871 the parish was made up of 2,805 "souls."[2]

Established in its own building, the parish organized its first fund-raising bazaar at Mechanics Hall, a center of public life in the city. After four days of sales, raffles, and games, the pastor totaled up the receipts and discovered a net profit of over $3,900. What he did next demonstrated that the French-Canadians' sense of their identity extended into places and issues with which the Assumptionists in Europe were already deeply involved. Having called for silence in the hall, the pastor informed his flock that he was about to do what he had promised God he would do if the profits of the bazaar should exceed $3,000. In the name of the parish he would send $400 — a tithe — to each of "two beings who are very dear to all of us who are Catholic and French-Canadian: Pius IX and France, both of whom are so unfortunate today."[3]

Fr. Primeau was alluding to the effects of France's military failure in a brief war with Prussia. A shocking Prussian victory at Sedan that November had precipitated a revolution in Paris that overthrew Napoleon III and the Second Empire. The resulting provisional government of France resolved to fight on, but by January 28 it had concluded that further resistance would be futile and signed an armistice. Just days before the bazaar in Mechanics Hall, the prostrate French had, under the provisions of the armistice, elected a government with authority to negotiate a definitive treaty of peace.

The Franco-Prussian War brought to a successful conclusion Chancellor Otto von Bismarck's long campaign to unify Germany under Prussian leadership. Unhappily for Pius IX, it also permitted the completion of a long unification campaign in Italy. In 1870, deprived of their French protectors, the Pope and his troops could do little to prevent the seizure of Rome by the kingdom of Italy. Refusing to concede the loss of the papacy's temporal power, Pius IX had become a self-proclaimed "prisoner in the Vatican."

Among the many Europeans brooding over the dual humiliations of France and of the papacy was Fr. Emmanuel d'Alzon, militant advocate of papal supremacy, sworn adversary of what he called "the Revolution," and founder of the Augustinians of the Assumption.

Emmanuel d'Alzon

Fr. Emmanuel d'Alzon, A.A.

Fr. d'Alzon had been born in 1810, during the Empire of Napoleon, into one of the wealthiest aristocratic families in southern France. The Bourbon monarchy was restored in 1814, and in 1823 Fr. d'Alzon's father, the Viscount Henri Daudé d'Alzon, went to Paris to serve in the Chamber of Deputies, bringing his son with him. In deference to his parents, the young student soon put aside notions about a military career and began the legal training that might lead to a career in diplomacy, administration, or the judiciary. More importantly, in Paris d'Alzon associated with young men eager to promote a revival of Catholicism in France, reversing the anti-Catholic tide that had flowed from the Revolution of 1789. They also wanted to oppose a powerful force within the Catholic Church in France, the long tradition of "Gallicanism," a sense of national religious autonomy, and, thus, of resistance to papal authority. On that question, young d'Alzon early and forever sided with the "ultramontanes," for whom the legitimate seat of all ecclesiastical authority was located "beyond the mountains," in Rome.[4]

Though he had already decided to devote his life to the defense of religion, in 1830 d'Alzon confided to a friend that he had also resolved to become a priest. It took him two more years to reveal to his parents the fact that their only son would lead a celibate life. In March 1832, d'Alzon slipped out of his parents' chateau at Lavagnac to enter the seminary of Montpellier. Though he readily submitted himself to the rigors of seminary life, he found himself intellectually and culturally unfulfilled by the provincial instruction at Montpellier and departed for Rome, where he arrived at the end of 1833.[5]

At the seat of Catholic governance, d'Alzon set out upon a fresh course of studies, living in a monastery and choosing his own professors, among them some of the most distinguished theologians of the era. He also developed a fresh passion for papal authority that would guide him for the rest of his life. Despite any reservations he may have had about Vatican politics and procedures, the young Frenchman had come to the spiritual conviction that "one must always work for Rome, sometimes without Rome, never against Rome." It was in Rome that he was ordained to the priesthood on December 26, 1834. In late 1835, he returned and offered his services to his bishop, who appointed the twenty-five-year-old aristocrat vicar-general, the second-ranking official in the diocese.[6]

Fr. d'Alzon would hold that office until his death in 1880. A man of prodigious energy, he threw himself into the battle he had long prepared to wage. He initiated or sponsored convents, orphanages, a refuge for "fallen" girls, projects for the education and moral development of the young, and associations to assist the poor. His zeal was powerfully reinforced by the personal wealth, inherited from his family, which he was quick to employ in support of his projects. In 1838, he began a correspondence with Eugénie Milleret, who became Mother Marie-Eugénie de Jésus, the founding superior of the Religious of the Assumption in 1839. The length and depth of their collaboration led the Congregation to regard him as a founder, though he had not officially carried out that function.[7] The daughters of Mother Marie-Eugénie would in time join the sons of Fr. d'Alzon in Worcester.

Fr. d'Alzon as Educator

Fr. d'Alzon's formal career as an educator began in the 1840s. While on vacation in July 1843, he had been surprised to learn that a priest colleague had, in both their names, carried out their tentative plan to purchase a property in Nîmes that housed a boarding school under the patronage of Our Lady of the Assumption. Fr. d'Alzon's plan was to close the school and turn the residence into a convent for Carmelite nuns. However, the schoolmaster who was his tenant at first refused to give up his lease, and by the time the men finally came to terms, in October 1844, a new school year was well under way.

With a functioning school on his hands, Fr. d'Alzon harkened back to a longstanding dream of shaping a new Catholic elite through a network of Catholic colleges and universities. Rather than close the school, he decided to revive the institution and use it as a means of combating the state law that required students to attend a public college for the last two years of their six-year secondary training — roughly comparable to the first two years of college studies in the United States. In Fr. d'Alzon's revived institution, religious instruction and overall discipline were entrusted to diocesan priests, while academic subjects were taught by a small group of university-trained professors, some of whom he convinced to give up their secure positions in Paris to teach at Assumption College in Nîmes.[8]

Boasting a faculty with credentials rivaling or surpassing those of professors in state institutions, Assumption College received positive evaluations from public education authorities, though the more advanced students were still required to take courses at the state-operated royal college of Nîmes. Fr. d'Alzon nevertheless made some political progress, gaining from the national ministry of education permission to teach some advanced courses within his own walls. He then went beyond what was officially permitted, keeping most of his older students at Assumption College and provoking fervid protests from local authorities.[9]

The education Fr. d'Alzon wanted to offer at Nîmes combined high academic quality with the formation in the souls of the students of a personal relationship with Jesus Christ.[10] He hoped to produce many candidates for the priesthood, and, after founding a new religious congregation in 1850, he had a special desire to see his graduates enter that community. While his school attracted the cream of the aristocracy and upper bourgeoisie of southern France, those young men of family and property did not imitate the aristocratic Fr. d'Alzon in turning their lives toward the priesthood. After decades of disappointment, at the close of the school year in 1870, the director announced, "Since so few of you are responding to the desire we had conceived to see you embrace the ecclesiastical state, we will address ourselves to the poor." Assumption College in Nîmes would live on, but priestly recruits would be sought elsewhere.[11]

Founding of the Assumptionists

In 1844, as he was preparing to revive the college in Nîmes, Fr. d'Alzon reflected on the ways ambitious clergymen could damage the Church, and at Turin he took a vow never to seek or accept elevation to "ecclesiastical dignities." With all such prospects behind him, he confided to Mother Marie-Eugénie that he was being drawn toward an idea he had considered a decade earlier, that he should become the founder of a religious community of men "to regenerate society." Mother Eugénie provided strong encouragement, in part because she had for some time been hoping for the establishment of a male congregation under the banner of the Assumption, a community dedicated, like hers, to teaching.[12]

Fr. d'Alzon decided to build his new Congregation from within the college of Nîmes. In October 1844, after he had preached a retreat to his professors, all eight — three priests, a deacon, and four laymen — joined him in forming the Assumption Association, out of which Fr. d'Alzon hoped he could shape both his new religious Congregation and a "Third Order," a society of priests and dedicated laymen who would not live in religious community but would take vows appropriate to their state. On Christmas Day 1845, Fr. d'Alzon, four other priests, and a layman entered a self-founded novitiate to begin the religious community. Two days later, a novitiate was established for the Third Order. For reasons of prudence, all of this was done privately and unofficially, neither the bishop nor the government being let in on what was happening.

The new Congregation had as its explicit goal to spread the kingdom of Jesus Christ, and it would pursue that aim through: (1) education; (2) the publication of educational books; (3) works of charity, through which children would be prepared to do their Christian duty and to reconcile the rich and poor classes; (4) the preaching of retreats; and (5) foreign missions and efforts for the destruction of heresy and schism. In the beginning, for Fr. d'Alzon, these terms referred to Protestant heresy and Anglican schism, but he later expanded the scope to include the Orthodox schism.

The founder had a difficult time holding his band together, due in part to the high standards he set, and also in part to the refusal of his bishop to authorize any public or official taking of vows. By the end of the school year in 1848, there were only three novices besides himself. A year later, when there were five, the bishop formally authorized a novitiate, but still would not allow public vows. Fr. d'Alzon made arrangements for private vows to be taken by himself and five others on Christmas Day 1850.

On Christmas Eve, the bishop suddenly gave permission for public vows. Before the Midnight Mass, Fr. d'Alzon solemnly pronounced his temporary vows of poverty, chastity, and obedience and received the vows of his colleagues. One year later, their novitiate over, Fr. d'Alzon and four others pronounced perpetual vows at the Midnight Mass before the students and faculty in the chapel of Assumption College. The professed religious put on a religious habit, a black cassock similar to those worn by diocesan priests, along with a monastic hood, Fr. d'Alzon's way of emphasizing that his new community would combine aspects of monastic life with active engagement in the world.[13]

Confronting the Foe

Engagement with the changing world of his time indeed characterized the remaining thirty years of Fr. d'Alzon's life. When the Falloux Law in 1850 provided for Catholic participation in the regulation of higher education, he reluctantly accepted appointment to the nation's Higher Council for Public Education, where he served until 1853, working to exempt Catholic higher education from supervision by the secular university. To rally Catholic educators he founded the R*evue de l'Enseignment Chrétien* (Review of Christian Education) in 1851. He also carried on a long campaign of education and preaching against Protestantism, especially in his diocese of Nîmes.[14]

Fr. d'Alzon, in the late 1850s and the 1860s, was also waging the battle over papal authority, both spiritual and temporal. Within the Church, he articulated a vigorous anti-Gallican, pro-Roman position, and he also came to the defense of the Pope when his temporal claims were under assault. His ultramontane campaigns exasperated Gallican churchmen to the point where a neighboring bishop stripped Fr. d'Alzon of priestly functions within his diocese. One consequence of that action was that Fr. d'Alzon was unable to administer the last sacraments to his mother as she lay dying in Montpellier in 1860.[15]

In 1860, Fr. d'Alzon accepted invitations to send some of his few Assumptionists to serve as missionaries to Australia and London. His missionary interest, however, was drawn even more powerfully to the East.[16] With the death of his mother and the retirement of his father, who would die in 1864, Fr. d'Alzon again came into control of large financial resources, and he dreamed of establishing an Assumptionist mission in Palestine. However, he was deflected by Vatican authorities to Bulgaria, then a part of the Turkish-based Ottoman Empire. Fr. d'Alzon sent some of his men there to combat the thousand-year-old Orthodox schism, and he sent women as well. In 1864, Fr. d'Alzon founded the Oblates of the Assumption, principally to assist the Assumptionists already at work in "the East."[17]

Late in 1869, as an official theologian, Fr. d'Alzon accompanied his bishop to Rome to take part in the First Vatican Council, an event that in many ways embodied Fr. d'Alzon's battle against the intellectual currents of the modern West and his devotion to papal authority. For nine months, Fr. d'Alzon played an active role in the behind-the-scenes business of the council, working against the delaying tactics of those who opposed the proclamation of a dogma of papal infallibility. In a letter to the Oblates of the Assumption, the founder encapsulated the lessons of the hour:

> *By attentively studying the purpose of the Council, one can see that what is being proposed is above all to restore the whole supernatural order brought to Earth by Our Lord. It is being attacked by incredulity, by naturalism, by rationalism, and by socialism. In opposition to incredulity, place the principles of the faith; in opposition to naturalism, place the whole plan of supernatural assistance and supernatural hopes; in opposition to rationalism, place the unshakable foundation of divine authority, strengthened by the doctrine of the infallibility of the Pope; in opposition to socialism, place the more perfect idea of social principles such as they are communicated by the great Christian society: the Church.*[18]

During the disastrous Franco-Prussian War of 1870, Fr. d'Alzon sent money and material to support priests of his congregation who served as volunteer chaplains in the French army, helped the Oblates of the Assumption function as military nurses, and preached daily in the cathedral of Nîmes at masses for the victims of the war. After the French defeat and the fall of Napoleon III, the new moderate republican government, in the Gallican tradition, asked Fr. d'Alzon to accept appointment as a bishop; for the third time in his career, he declined the honor.[19]

In time, Fr. d'Alzon and his followers, more than any other group, would shape the French Catholic response to the dual disasters of 1870, and to the equally frightening emergence of the revolutionary Paris Commune, suppressed in a bloodbath in 1871.[20] They interpreted these events as divine punishment for France's abandonment of its character as a Catholic nation, a trail of sins that led back through the French Revolution of 1789 to the impious rationalism of the Enlightenment and the heretical rebellion of the Reformation. What the situation called for was expiation, repentance, and atonement, a spirit that manifested itself during the ensuing decades in a variety of popular devotions, many led by Assumptionists, in particular pilgrimages to sacred shrines.[21]

The political abasement of the formerly Christian French nation and the defeat of the papal power were also seen as manifestations of Satan's triumphs in a sinful world. The "Revolution" that stained the modern world was one with the rebellion of Satan and his followers against the authority of God. The Assumptionists were particularly concerned about the deeply antireligious spirit and the growing political power of freemasonry in France. Regeneration, they believed, would require a militant Catholicism, fully attuned to the supernatural character of the war that was under way and fully committed to waging it to the death.[22]

The Alumnates

In early 1871, while the French-Canadians in Worcester were sending their tithes to Pius IX and to France, Fr. d'Alzon was deeply engaged in the creation of the alumnate, a new form of educational institution; one he hoped would shape the sort of warrior-priest that the crisis of the age demanded. The founder was told about an old sanctuary dedicated to the Blessed Virgin that stood

The first alumnate, Notre Dame des Chateaux

on an isolated knoll near Beaufort, facing Mont Blanc. Adjacent to the sanctuary could be found the ruins of an old seigneurial residence, with only the towers and the dungeon still intact. With characteristic boldness, Fr. d'Alzon in April 1871 visited the site, which could be reached only on foot or by mule, ordered the necessary repairs, and settled on August 1871 for the opening of Notre Dame des Chateaux, the first alumnate.[23]

Fr. d'Alzon drew the word "alumnate" from a term used to designate a medieval institution, a school attached to a monastery, where boys aspiring to the priesthood were received. The school would be exclusively dedicated to the ecclesiastical education of poor children, who would, at the conclusion of their classical studies, be free to join the diocesan clergy, the missions, or a religious community. With some exceptions, boys would have to be at least twelve years old when entering and literate enough in French to begin the study of Latin. Classical studies, normally spread out over six years, would be covered in five, after which students would proceed to two years of philosophy at a diocesan seminary or religious scholasticate.

The atmosphere was in some ways to resemble that of a monastery, with the boys reciting the Divine Office at designated hours of the day, along with the monks who were their teachers and constant companions. The students and religious would do all chores, including manual labor in the fields. Except for kitchen help, there would be no servants. During vacations, there would be no returning to families and thus no dangerous exposure to worldly temptations. During recreation periods, the religious were to join in the students' games, and discipline was to be based on honor and emulation rather than punishment.

Above all, the alumnates were to form militant priests, "warriors against Satan and the world," rigorously prepared to engage in the great combat between Heaven and Hell. "We want," Fr. d'Alzon wrote, "to make the blood of Jesus Christ flow with an even greater fullness in their souls, and to do that by preoccupying them above all with the love of God, with his Son, with the Blessed Virgin, with the Church, with the saints, in a word, with that great cause in which the interests of Heaven are engaged against those of Hell."[24]

Fr. Vincent de Paul Bailly, A.A.

Pilgrimages and the Press

In another field of activity Fr. d'Alzon and his Assumptionists were momentously successful. In 1872, in Paris, Fr. François Picard formed the Association of Our Lady of Salvation. Originally intended to encourage public manifestations of faith among workers, the effort turned into a series of popular pilgrimages led by Fr. d'Alzon and other Assumptionists, especially Fr. Vincent de Paul Bailly, to local, national, and international shrines. Between 1870 and 1878, for example, the

Assumptionists led almost a thousand pilgrimages to Lourdes, involving more than half a million pilgrims.[25]

Out of the pilgrimage movement grew another activity that would have even greater historical consequences, the founding of a popular Assumptionist press. In July 1873, the Congregation published the first issue of *Le Pèlerin* (The Pilgrim). It began as a tool for organizing pilgrimages but grew into a popular illustrated weekly after coming under the direction of Fr. Vincent de Paul Bailly in 1877. By 1883 the publication had over 45,000 subscribers and would reach a peak of 196,000 by 1900. It became the major source of financing for the Assumptionist publishing house, the Maison de la Bonne Presse. [26]

In 1880, when the religious congregations of France were coming under pressure from government secularizers, Fr. d'Alzon decided to establish a newspaper aimed directly at anticlerical forces, especially the freemasons and their lodges. The vehicle was *La Croix* (The Cross), which emerged first in March 1880 as a review aimed at the educated classes, and would become a popular daily newspaper in 1883.[27]

Attack, Death, and Dispersal

The appearance of *La Croix* in 1880 virtually coincided with the publication of two anticlerical decrees by the government of France. The Jesuits were ordered to leave the country, and all of the congregations that had not been authorized by the state — the vast majority of both male and female religious communities in France — were ordered to seek authorization within three months or disband. The religious congregations, with the support of most bishops, rallied around the Jesuits and agreed not to submit their constitutions for scrutiny and approval by the national government, which seemed prepared to use the police to remove all unauthorized priests, brothers, and nuns from their homes.[28]

The Jesuits were expelled, but the government was reluctant in the face of the massive resistance to carry through on the rest of its threat. It indicated through intermediaries that the closing down of the communities might be delayed, pending further legislation, but only if the congregations first made some show of loyalty to the republic. Accordingly, a loyal "Declaration of the French Bishops" was secretly formulated, and Pope Leo XIII, without issuing an order, let it be known that he wanted the leaders of all the nonauthorized congregations to sign, for the good of the Church. Fr. d'Alzon, who had strongly opposed any submission to the law, immediately submitted to the will of the new pope, fell to his knees, and signed the declaration. To those who expressed surprise, he remarked, with characteristic militancy, "A colonel who disobeys his commander on the field of battle deserves to be shot."[29]

While the declaration was being secretly circulated for signatures, the text appeared in a newspaper. In the ensuing uproar, the compromising French government fell from power. It was replaced by one led by the anticlerical Jules Ferry, who put the decrees into execution. In Paris and in Sèvres, Assumptionists were expelled from their residences on November 5, 1880. At the headquarters in Nîmes, things took longer, because there Emmanuel d'Alzon lingered on his deathbed. Only after his death on November 21 and his funeral a few days later, did the authorities move in and disperse the community.[30]

Thirty years after the founding of the Assumptionists, Fr. d'Alzon left behind a scattered community of sixty-five religious. Only twenty-seven of them were priests, the others falling into the categories of choir brothers — those preparing for the priesthood — or lay brothers, who did not aspire to ordination. Despite their small numbers, they had already achieved considerable influence in France, and with the help of their new newspaper they would become a major political force in the decades after Fr. d'Alzon's death.

The 1880 suppression of the religious congregations turned out to be temporary. It was difficult under any circumstances to prevent people from living together, especially if they did so discreetly and in small numbers. In addition, the vacillating politics of successive governments made the decrees even less likely to prevail in the long run. The man Fr. d'Alzon had chosen to succeed him as superior general, Fr. François Picard, slipped back into the Assumptionist residence on Rue François Ier in Paris only days after his expulsion, and from there he directed the Assumptionists for the next twenty years.[31]

The Dissolution of the Assumptionists

At the end of those two decades, the Assumptionists would face a much more focused assault from the national government and a dispersal that would send some members of the congregation to North America and to Worcester. Of the great diversity of works undertaken by the enterprising Congregation, the one that finally landed them in serious political trouble was the publication of *La Croix*. At first an innovative experiment, it became the largest and most powerful Catholic journal in France, and one of its largest daily newspapers. By March 1893, there were nearly 2,000 subscription committees working in various localities, and by 1895 almost 100 regional versions of the newspaper were being published, most of them weeklies but some appeared daily.[32]

Though Emmanuel d'Alzon had always insisted that the interests of the Church could not be attached to monarchy or to republicanism or to any particular form of government, the implacable hostility of *La Croix* to the republican governments of the era could readily be interpreted as hostility to republicanism itself. The newspaper's criticisms were always expressed in terms of religion, but it was a religion strongly tinged with French nationalism. For the 1889 election, for example, it created among voters the Ave Maria League, endowing it with a stinging electoral platform:

> *No more thieves. No more laicizers. No more persecutors. No more Freemasons. No more Jews. No more Prussians. No more foreigners to govern France. Nothing but honest men. Nothing but Catholics. Nothing but Frenchmen.*[33]

Pope Leo XIII, in February 1892, inaugurated a policy known as the *Ralliement*, directing French Catholics to accept the republic and to work together for the good of the Church within the republican system. *La Croix* reluctantly submitted to the papal directive in an editorial signed by the Assumptionist superior general, Fr. François Picard.[34]

As it curtailed some of its antirepublican rhetoric, *La Croix* increasingly blended nationalism with its version of a wave of anti-Semitism that was surging through Europe late in the nineteenth century. In France the revival of anti-Jewish propaganda was marked by the publication of Edouard Drumont's *La France Juive* (Jewish France) in 1886.[35] The Assumptionists had long pointed to freemasonry as the Satanic center of antireligion; now *La Croix* began to argue that behind the masons were the forces of international Jewry: "Freemasonry" it declared, "has been organized by Jewry, to subvert Christian nations, to ruin them, and to enslave them."[36]

Such beliefs helped to make *La Croix* one of the most virulent and conspicuous voices in the divisive campaign to uphold the 1894 conviction of Captain Alfred Dreyfus. A young Jewish army officer, Dreyfus had been arrested, charged with passing military information to the Germans, and found guilty by a court-martial. As evidence tending to exonerate Dreyfus became public, the French nation became deeply divided over the case. The political right — including *La Croix* — besides appealing to anti-Semitism, preached that a citizen had a moral duty to support the army since it was the only nonpolitical symbol of the nation. The political left insisted that the individual rights of the accused took precedence and raised alarm over the anti-Dreyfus collaboration of intransigent Catholics, militant anti-Semites, and inveterate royalists. When a military tribunal on September 9, 1899, upheld the original conviction of Dreyfus, *La Croix* found its arguments and prejudices once again vindicated.[37]

One of the highest priorities of President Pierre Marie René Waldeck-Rousseau was to find a way to cool the long-burning

Dreyfus controversy. His emissaries persuaded the convicted officer to withdraw his appeal to the nation's highest court and to accept instead a presidential pardon, which was issued on September 19.[38] In his drive to restore peace and order, Waldeck-Rousseau then turned his attention to the Assumptionists. In addition to their notorious attacks on Dreyfus and on successive republican governments, the Congregation had, through their leadership of the *Comité Justice-Égalité*, been engaged in effective nationwide electoral organizing and campaigning. In the elections of 1898 they had actively supported candidates known to be hostile to the republic itself.[39]

On November 11, 1899, at eight o'clock in the morning, police officers invaded the fifteen houses occupied by the Assumptionists in France. Over 120 government officials swarmed over the headquarters on Rue François Ier in Paris, seizing money, documents, and other evidence. On January 22, 1900, the twelve Assumptionists who had been living there appeared in court, charged with acting as a political association under the guise of a religious community. On January 24 the twelve were fined 16 francs each, but much more importantly, the court declared "dissolved the association known as the Congregation of the Augustinian Fathers of the Assumption." To function openly as a religious community, the Assumptionists would have to leave France.[40]

The defendants appealed, if only to gain time. When protests erupted from the Catholic press and from some members of the hierarchy, the government responded by depriving protesting priests, bishops, even the Cardinal of Paris, of their state stipends.[41] The Waldeck-Rousseau government indicated to Pope Leo XIII that if the Assumptionists were quietly sacrificed no other congregation would be endangered. In a letter to the Bishop of Orleans, the pope made it clear that, despite his sympathy toward the Assumptionists, he considered any politically oriented protest to be dangerous for the Church in France.[42]

The court of appeals on March 6 sustained the verdict and the punishments declared by the trial court. Although the Assumptionists had until the end of June to disband, on March 17 the pope sent an oral order to Fr. Emmanuel Bailly, brother of the editor of *La Croix*. The Assumptionists were to cease forthwith their association with the newspaper, though they could as individuals continue to be connected with other publications of La Bonne Presse. In the d'Alzonian spirit of submission to Rome, the four priests who had shared editorial duties at the controversial newspaper gave up their work, the office of *La Croix* was transferred from the Congregation's Parisian mother house to the printing establishment of La Bonne Presse, and editorial control was turned over to laymen.[43]

Unable to function openly as a religious community, many Assumptionists remained in France, living alone or in very small groups. They and other religious congregations continued to suffer the consequences of official anticlericalism, which culminated in a December 1905 law voiding the 1801 Napoleonic Concordat with the Vatican. The new law separated church from state, and brought to an end centuries of government funding of the Church.[44]

Assumptionists in the United States

Pulled by opportunities for missionary activity and pushed by the threat of political attack in France, Assumptionists had been taking up posts in several countries prior to the governmental crackdown of 1899. The Congregation's first step toward the United States was unknowingly taken in 1860 by Fr. Henri Brun, who was assigned to London as chaplain to the second convent opened there by the Religious of the Assumption. During his two years in the British capital, Fr. Brun learned English and explored the possibility of starting an Assumptionist mission there. In 1862, he was granted his wish to be assigned to the mission in Australia, where he served until 1873, further polishing his English and otherwise gaining experience that would later serve him well in the United States.[45]

Fr. Henri Brun, A.A.

During a trip to Europe in 1890, the archbishop of New York, Michael A. Corrigan, became aware of the work carried out by the Little Sisters of the Assumption, a community founded in 1865 by Antoinette Fage and Étienne Pernet, an Assumptionist priest. The Little Sisters worked among the poor, of which New York City had its full share. At Corrigan's request, five sisters left Paris for New York on April 10, 1891, accompanied by a chaplain newly assigned to them, Fr. Henri Brun. The Assumptionist presence in North America thus dates from Fr. Brun's arrival at the French Line pier at the foot of West 14th Street on April 19. The first representative of the Congregation in the New World had participated in the birth of the Congregation when he pronounced religious vows with Emmanuel d'Alzon on Christmas Eve in 1850.[46]

New York remained the base of early Assumptionist activities in the United States, with the exception of a disastrous attempt to establish a mission among French-speaking African-Americans in Louisiana. Launched in 1892 and abandoned in 1900, the Louisiana mission suffered from disagreements between Assumptionists and diocesan authorities, self-defeating behavior by some of the few Assumptionists on the scene, and resistance from Louisianians, black and white, lay and clerical.

In New York, Fr. Henri Brun concluded that promising opportunities were available to the Assumptionists, but before any collaborators were assigned to assist him, he died on January 15, 1895. His immediate successors, who tended to regard their "exile" in New York as temporary, made little progress in establishing solid foundations. In 1897, two of them opened an alumnate, despite diocesan warnings that it would have to be closed when a diocesan minor seminary was established. The alumnate closed in 1900, the same year the Louisiana mission was abandoned.

A new avenue opened in 1901 when Archbishop Corrigan took the Assumptionists up on their offer, made a few years before, to minister to the Spanish-speaking people of New York. Thanks to the opening of an Assumptionist novitiate in Spain and of a mission in Chile, Spanish-speaking personnel would be available for the new task. Early in 1902, the six Assumptionists in New York opened a modest chapel on West 14th Street. In time the parishes of Our Lady of Guadalupe and Our Lady of Esperanza (on 156th Street) would become major Assumptionist responsibilities.

Fr. Thomas Darbois, A.A.

The Spanish-speaking apostolate in New York began under the leadership of Fr. Thomas Darbois, who had spent ten years in Chile. As that effort was getting under way, a new opportunity presented itself, one that would, in a convoluted fashion, move the Assumptionists further in the direction of Worcester. Early in 1902, Fr. Charles Crevier, pastor of the French-speaking Precious Blood Parish in Holyoke, Massachusetts, dined in New York with Fr. Darbois and his colleagues. As Assumptionists often did, they conveyed to their guest their enthusiasm for alumnates. To their surprise, Fr. Crevier suggested that they initiate such an enterprise in the town of Granby, Massachusetts, where he owned a farm and a house, both of which he would place at their disposal. Unable to act on such an offer on his own authority, Fr. Darbois provided the Holyoke pastor, who was preparing to leave for Europe, with letters of introduction to his Assumptionist superiors.

After his return, Fr. Crevier drafted a formal proposal for the Assumptionists to establish an industrial, commercial, and agricultural school for French-speaking boys in Granby, in the Diocese of Springfield. Noting that he was making his offer on June 24, the feast of John the Baptist, patron saint of French Canada, Fr. Crevier

Fr. Charles Crevier

stressed the ethnic bonds linking France and Quebec. He also offered Granby to the Assumptionists as a refuge from their political distress. "The children of Canadians living in the United States of America are asking the Very Reverend Father General of the Reverend Fathers of the Assumption to send his sons to their assistance," he wrote. "While the French Republic is expelling you from your country the American Republic gladly receives you. What a lesson for France."[47]

Warned of Fr. Crevier's reputation for changing his mind, Fr. Darbois forwarded the proposal to the superior general, cautioning him not to assign personnel until the land transfer had been sealed. Brushing aside the warning, Fr. François Picard approved the proposal in principle and began selecting priests and brothers to send to Massachusetts. Fr. Crevier wanted an immediate start, but delays resulted from the hesitation of both Springfield Bishop Thomas M. Beaven and the Assumptionists. Negotiations extended into April 1903, when Fr. Darbois signed a contract to establish the new school. On April 27, two Assumptionist priests and two lay brothers arrived in New York from France. The designated superior for Granby, Fr. Antoine Silberman, proceeded immediately to the scene, which by the end of May had an Assumptionist community of four priests and three lay brothers.[48]

They found themselves living in cramped quarters in Fr. Crevier's farmhouse, with the Holyoke pastor in residence and apparently believing that he was to be in charge. Fr. Crevier had reserved the four best rooms for himself and his nephew and had designated two others for servants, leaving the Assumptionists with two small bedrooms. A succession of new demands from the Holyoke pastor led to a succession of complaints by the Assumptionist superior. In late May, Fr. Silberman expressed to his superior general his regret "that no competent authority examined this project before five religious were sent from France."[49] A few years later, looking back on those weeks of conflict, an Assumptionist described Fr. Crevier of Holyoke as "a whimsical and changeable man," but also observed that "Fr. Antoine [Silberman] found everything bad and thought he would render a service to the Congregation by making this foundation fail, to which he had been sent from the East."[50]

Responding, perhaps, to Fr. Silberman's pessimistic reports, on June 15, 1903, Emmanuel Bailly — superior general after the April 16 death of François Picard — once again placed full authority over the Granby project in the hands of the superior in New York, Fr. Thomas Darbois. As he would later demonstrate in Worcester, Fr. Darbois believed that the Assumptionists should be willing to make heroic sacrifices to get themselves established in the Springfield diocese and begin the work of cultivating priestly vocations.[51] With Fr. Darbois pushing him forward, the reluctant Fr. Silberman, in a letter to Bishop Beaven, sighed, "My superiors having told me to begin, I will do it. May God help me in the midst of the many difficulties I anticipate."[52]

Bishop Thomas M. Beaven

From the Assumptionist point of view, the most serious of their many problems in Granby was Fr. Crevier's refusal to live up to his promises, especially his promise to execute a deed to the property. The breaking point was reached on July 4, 1903, when the Assumptionists hastily packed their bags and departed. "Fr. Crevier has thrown us out of Granby," Fr. Darbois reported to Bishop Beaven, "ordering us to leave immediately because we cost too much to feed."[53] Though they

Rev. Joseph Brouillet

went to Springfield specifically to see the bishop, Frs. Darbois and Silberman found him unavailable. After hearing the story from an aide, Bishop Beaven wrote sympathetically in his journal that the Assumptionists "took up their abode in good faith," had received a deed from Fr. Crevier, had returned it to him when he asked to look it over, and had been unable to get it back. "The fathers," he wrote, "in a truly religious spirit declared that they wl'd [would] make legally no demand for its return or the fulfillment of the contract. Such a procedure would be injurious to religion and they preferred to abandon the work."[56]

Wasting no time, Fr. Darbois in a July 5 letter asked Bishop Beaven to allow the Assumptionists another post in the diocese. "Along with some work of general interest," he wrote, "we could have our apostolic school [alumnate] and several priests available for missionary work and to help the pastors." The next day the bishop replied,"I feel very keenly the situation at Granby, and regret exceedingly the outcome. I hope in a few days to give you an answer in relation to your request."

Bishop Beaven did not in fact give the Assumptionists permission to found another house in his diocese until early September 1903.[57] Possibly at the bishop's suggestion, Fr. Darbois on September 3, left New York for Worcester and the welcoming rectory of Fr. Joseph Brouillet, pastor of the same parish of Notre Dame des Canadiens that had sent $400 to the pope and $400 to France three decades earlier.[58] By this time, the French-Canadian population made up around 20,000 of the 130,000 residents of Worcester. They were organized into three parishes, each with an elementary school conducted by the Sisters of St. Anne. The community supported a daily French newspaper and an annual directory of the French-Canadian population. While Fr. Darbois, who was still superior in New York, helped minister to Notre Dame parish, the pastor, Fr. Brouillet, helped the Assumptionist leader search for property that could be used as a residence for missionaries and, in time, as a school where young men could begin training for the priesthood.

Fr. Brouillet was especially impressed with the possibilities of a house and barn on a six-acre hillside in a section of northern Worcester known as Greendale. On September 22, Fr. Darbois reported to his superior general that he would be buying that property within a few days for $10,000. He described Greendale as a suburb without a Catholic church, within eight or nine kilometers of the large industrial city of Worcester, on the opposite side of the city from the Jesuit-conducted College of the Holy Cross.[59]

Three days later, Bishop Beaven notified the priests who served as his diocesan "consulters" that he had granted the Assumptionists permission to purchase a house and land at Greendale, "where they may establish a Mission House, or place of residence for those who may be engaged in giving Missions to French-Canadian parishes."[60]

Fr. Darbois that same day wrote to his superior general, Emmanuel Bailly, to notify him of the bishop's consent and of his plan to purchase the Greendale property the following day. Somehow this transaction was called off or retroactively canceled, apparently through the intervention of the superior general. Caution about finances and staffing were undoubtedly involved, but Fr. Bailly also insisted that Fr. Darbois secure written permission from both the Congregation and the bishop before moving forward.[61]

On October 5, still living at the Notre Dame rectory, Fr. Darbois notified the bishop that he had brought a second Assumptionist, Fr. Isidore Gayraud, to join him and to give a retreat at St. Anne's Orphanage, conducted by the Gray Nuns. He added that he had not yet bought any property. The Assumptionist Congregation was poor and persecuted, he explained, and besides, he had yet to receive written permission from the bishop.[62] Through a subordinate, Bishop Beaven immediately sent a message stating that the Assumptionists were free to settle in Greendale without written permission from him.[63]

On November 1, Fr. Darbois notified Fr. Bailly that he had selected another house in Greendale, a 10-room building on a 35 x 120-meter lot on Fales Street. "That," he wrote, "will be our missionaries' residence. By our work we will be able to pay for the house rapidly. Then we will be able to build the alumnate, a work the Bishop authorized in Granby, and orally for Worcester. Yesterday, when he heard we were buying a house, he sent me a letter indicating some restrictions."[64] Fr. Darbois' concern about restrictions was apparently prompted by an October 29 letter in which the bishop withheld approval of the educational dimensions of the project. For the time being, Bishop Beaven wrote, a house "for Missions" was all that was practical; when circumstances changed it would be time enough to consider "the prospective works of the community," including an apostolic school or a commercial or industrial college.[65]

On November 5, Fr. Darbois reported that he had purchased the house at 27 Fales Street for $6,500.[66] The first religious moved in a week later, establishing November 12, 1903, as the date on which the Assumptionist foundation in Worcester began.[67] It was just under fifty-three years since Fr. d'Alzon had founded the Augustinians of the Assumption, nearly thirty-three years since Fr. Primeau and his parishioners had reached out to help the Pope and France. The first mass in the new foundation was celebrated the following day, November 13, 1903.[68]

The war against Satan and "the Revolution" could not be waged in the twentieth-century American milieu as it had been during the nineteenth century in France. In the decades ahead, much of the drama of the Assumptionist struggle in Worcester would stem from efforts of Fr. d'Alzon's sons to carry on their struggle for the coming of God's kingdom while learning to understand and to adapt to the new environment, a set of conditions that was itself always evolving.

For many purposes, at least at the start, the traditional terms would continue to serve, as they did for Fr. Darbois when he recounted the events of November 21, 1903, the day Fr. Joseph Brouillet, the pastor of Notre Dame, was scheduled to bless the new dwelling and its chapel.[69] As Fr. Darbois told the story, the priest charged with early morning preparations for the ceremony had some difficulty getting the fire in the furnace to burn well, so, to the smoldering wood and coal, he added a can of kerosene. The resulting blast sent a massive flame shooting out the furnace door, threatening to ignite not only the dry firewood piled nearby but the entire wooden structure. The threat, however, was brief. The flames damaged nothing more valuable than a few whiskers on the beard of the startled caretaker before settling back into the furnace, consuming the wood and coal, and warming up the building.

Fr. Darbois drew the moral: "The devil feared the good that would be done in the nascent school and wanted to destroy it at the outset. Providence foiled his malice."[70]

As they shivered through their first New England winter, the members of the tiny Worcester foundation undoubtedly prayed for divine assistance to rescue them from the many diabolical snares they expected to encounter in the months and years to come.

Notes

ABBREVIATIONS
AC — Archives of Assumption College, Worcester, MA
AG — Archives of the Assumptionist General House, Rome
AP — Archives of the Assumptionist Provincial House, Brighton, MA
PA — Archives of the Assumptionist Province of Paris
SP — Archives of the Diocese of Springfield, MA

1. Roy Rosenzweig, *Eight Hours for What We Will: Workers and Leisure in an Industrial City, 1870–1920* (Cambridge, England and New York, 1983), 12–13.
2. T.A. Chandonnet, *Notre-Dame-des-Canadiens et les Canadiens aux Etats Unis* (Montreal, 1872), 9–10, 31, 64.
3. Chandonnet, *Notre-Dame*, 44.
4. Polyeucte Guissard, A.A., *Un Siècle d'histoire Assomptioniste* (Worcester, 1950), 7–9.
5. Guissard, *Un Siècle*, 9–10; Siméon Vailhé, A.A., *Vie du P. Emmanuel d'Alzon* (2 vols., Paris, 1926 and 1934), I, 131; Monval, Jean, *Les Assomptionnistes* (Paris, 1939), 18.
6. Guissard, *Un Siecle*, 11; quotation from Vailhé, *Vie du P. Emmanuel d'Alzon*, I, 182.
7. Guissard, *Un Siècle*, 12–15, 20; Bernoville, *Un Promoteur*, 81–85.
8. Siméon Vailhé, A.A. *Vie du P. Emmanuel d'Alzon*, I, 328–39.
9. Louis Secondy, "Aux origines de la maison de l'Assomption à Nîmes (1844–1853)," in René Rémond and Emile Poulat, eds. *Emmanuel d'Alzon dans la société et l'Eglise du xixe siècle* (Paris, 1982), 242–43.
10. Monval, *Les Assomptionnistes*, 31–32.
11. Secondy, "Aux origines," 251–52; quotation from Guissard, *Un Siècle*, 68.
12. Bernoville, *Un promoteur*, 95–96; quotation on p. 96.
13. Bernoville, *Un promoteur*, 95–118.
14. Pierre Pierrard, "Le Père d'Alzon et la liberté d'enseignement en France," in Remond and Poulat, eds., *Emmanuel d'Alzon dans la societe et l'Église*, 111; Bernoville, *Un promoteur*, 116, 125–26.
15. Bernoville,*Un promoteur*, 146.
16. Bernoville, *Un promoteur*, 152–53.
17. Bernoville, *Un promoteur*, 152–53; Vailhé, *Vie du Père Emmanuel d'Alzon*, II, 355–65; Étienne Fouilloux, "L'úuvre orientale du Père d'Alzon vue par ses fils (1880–1980), in Rémond and Poulat, eds., *Emmanuel d'Alzon dans la sociètè et l'Église*, 208–9.

18. Bernoville, *Un promoteur,* 181–89; quotation from 181.

19. Bernoville, *Un promoteur,* 190–92.

20. René Rémond, *The Right Wing in France, 1815 to de Gaulle,* 2d. ed., (Philadelphia, 1969), 184.

21. Justin Mather, "The Assumptionist Response to Secularization, 1870–1900," in R. Bezucha, ed., *Modern European Social History* (Lexington, Mass., [1972]), 63–65.

22. Rémond, *The Right Wing in France from 1815 to de Gaulle,* 184–88.

23. Guissard, *Un Siècle,* 68–70.

24. Marie-Clément Staub, *Les Augustins de l'Assomption* (Bergerville, Quebec, 1926), 19; Guissard, *Un Siècle,* 71–79.

25. Bernoville, *Un promoteur,* 203–6; Justin Mather, "The Assumptionist Response to Secularization, 1870–1900," 64–65.

26. Bernoville, *Un promoteur,* 206; David M. Gallo, A.A., "Response to the Revolutions: Assumptionist Vision and Challenge in *La Croix,*" *Augustinian Heritage* 32, No. 1 (1986): 101.

27. Raymond J. Marion, "*La Croix* and the *Ralliement,*" (Ph.D. Diss., Clark University, 1957), 33.

28. Bernoville, *Un promoteur,* 220; Marion, "*La Croix* and the *Ralliement,*" 48; Monval, *Les Assomptionnistes,* 85.

29. Monval, *Les Assomptionnistes,* 86; Dansette, *Religious History of Modern France,* II, 45; Bernoville, *Un promoteur,* 221–22.

30. Monval, *Les Assomptionnistes,* 87–90.

31. Guissard, *Un Siècle,* 44.

32. Mather, "Assumptionist Response," 76–77.

33. Mather, "Assumptionist Response," 70.

34. Marion, "*La Croix* and the *Ralliement,*" 68.

35. Marion, "*La Croix* and the *Ralliement,*" 152.

36. L'Ami (pseud), "Mauvais quart d'heure," *La Croix,* August 11, 1894, quoted in Marion, "*La Croix* and the *Ralliement,*" 149.

37. Marion, "*La Croix* and the *Ralliement,*" 146–48, 158–60.

38. Malcolm O. Partin, *Waldeck-Rousseau, Combes, and the Church: The Politics of Anti-clericalism, 1899–1905* (Durham, N.C, 1969), 6–17.

39. Marion, "*La Croix* and the *Ralliement,*" 130–41; Partin, *Waldeck-Rousseau, Combes, and the Church,* 22

40. Marion, "*La Croix* and the *Ralliement,*" 166; Monval, *Les Assomptionnistes,* 132–33.

41. Marion, "*La Croix* and the *Ralliement*," 168.

42. Marion, "*La Croix* and the *Ralliement*," 168; Monval, *Les Assomptionnistes*, 130, 134.

43. Monval, *Les Assomptionnistes*, 135–36.

44. Monval, *Les Assomptionnistes*, 136–42.

45. Henry Moquin, A.A. and Richard Richards, A.A., *Assumptionists in the United States* (Worcester, 1994), 5–6. The following paragraphs are based on their account, 7–35.

46. Moquin and Richards, *Assumptionists in the United States*, 19.

47. Crevier to Thomas Darbois, June 24, 1902, AG 2OS-1.

48. Moquin and Richards, *Assumptionists in the United States*, 35–37.

49. Moquin and Richards, *Assumptionists in the United States*, 37. Their translation.

50. "Quelques notes sur la fondation de Greendale," appendix to Ernest Baudouy, "Carte de la Visite," April 27, 1908, 24, AG 2PK-19.

51. Moquin and Richards, *Assumptionists in the United States*, 37.

52. Silberman to Beaven, June 23, 1903, bound letter volume [1903–7], 59, SP.

53. Darbois to Beaven, July 5, 1903, bound letter volume [1903–7], 73, SP.

54. Beaven journal entry, July 4, 1903, "Records, Diocese of Springfield, 1871 to 1909," bound volume 488–89, SP.

55. Darbois to Beaven, July 5, 1903, bound letter volume [1903–7], 74, SP.

56. Beaven to Darbois, July 6, 1903, bound letter volume [1903–7], 72, SP.

57. Darbois to Emmmanuel Bailly, November 13, 1903, AG 2OS-27. Fr. Darbois wrote that "permission of Bishop Beaven for this foundation arrived within the octave of Our Lady of Consolation," a feast celebrated by Augustinians on September 4. (I am grateful to Fr. Wilfrid J. Dufault, A.A. for supplying the date of the Augustinian observance.)

    An account written in 1908 states that Bishop Beaven "around the middle of September . . . made it known that he considered us still admitted to his diocese and that he authorized us to establish ourselves in Worcester, to which Fr. Brouillet, Canadian pastor of Notre Dame, was calling us." ("Quelques notes sur la fondation de Greendale," appendix to Ernest Baudouy, "Carte de la Visite," April 27, 1908, AG 2PK-19.)

58. Early Assumptionist sources differ on Bishop Beaven's role, if any, in bringing Fr. Darbois into contact with Fr. Brouillet and on the date of Fr. Darbois' arrival in Worcester. An undated "Origines de Granby et de Greendale" (AC) claims that even before the Assumptionists left Granby, Bishop Beaven had "found the conduct of Fr. Crevier so unjust that as compensation he permitted us to choose an installation elsewhere and pointed out as a help to us Fr. Brouilett [who was] strongly desirous of making a Canadian foundation." According to

that document, Fr. Darbois proceeded to Worcester immediately after July 4. The 1908 "Quelques notes sur la fondation de Greendale" notes only that Bishop Beaven in mid-September approved an Assumptionist establishment in Worcester, "to which Fr. Brouillet, Canadian pastor of Notre Dame, was calling us." Fr. Omer Rochain's 1912 "Rapport sur la maison de Worcester" says that in August 1903 Bishop Beaven offered Fr. Darbois the opportunity to choose a location in Worcester and that Fr. Brouillet showed hospitality and encouragement (AG IH-7). Fr. Darbois' account published in 1925 states that the Assumptionists — who moved into their own residence in November — had enjoyed the hospitality of Fr. Brouillet's rectory during several months. ("Souvenirs de fondation pour servir a l'histoire de notre collège," *Almanach de l'Assumption* [Dec., 1925], 58.) The most recent account states that "throughout the long weeks of July 1903, the Assumptionists waited in vain for word from the Bishop" and does not give a date for Fr. Darbois'arrival in Worcester. (Moquin and Richards, *Assumptionists in the United States*, 1994, 45.)

59. Darbois, "Souvenirs de fondation pour servir a l'histoire de notre collège," *Almanach de l'Assumption* [Dec., 1925]; Darbois to Bailly, Sept. 22, 1903, AG 2OS-18.

60. Beaven to consulters, Sept. 25, 1903, bound letter volume [1903–7],178, SP.

61. Moquin and Richards, *Assumptionists in North America,* 45.

62. Darbois to Beaven, bound letter volume [1903–7], 200, SP.

63. M.A. Kelly to Darbois, Oct. 8, 1903, bound letter volume [1903–7], 201, SP.

64. Quoted in Moquin and Richards, *Assumptionists in the United States*, 46. Their translation.

65. Beaven to Darbois, Oct. 29, 1903, bound letter volume [1903–7], 23, SP. [Copy attached to Rochain, "Rapport sur la maison de Worcester (Etats Unis)," Sept. 27, 1912, AC.]

66. Moquin and Richards, *Assumptionists in the United States*, 46. Of that sum, $1,500 was paid in cash, $4,500 was borrowed from a bank, and $500 was borrowed from an individual identified only as "Monsieur Beaudry." (Darbois, "Etat financier de la maison do Worcester[Mass]," April 8, 1905, AG 2PK-17.)

In a 1925 article Fr. Darbois wrote that, on an unspecified date, Bishop Beaven sent a letter to the Assumptionists telling them that if they had not left the Notre Dame rectory and set themselves up independently within fifteen days, "the diocese would be closed to them." According to Fr. Darbois the Assumptionists, pushed by this ultimatum, returned to Greendale and discovered of a "For Sale" sign lying hidden in the grass outside a Fales Street building Fr. Darbois had admired during previous visits. The purchase was made, and "only a few days after the threatening letter of Bishop Beaven we were in our own dwelling." ("Souvenirs de fondation," 59.)

The Darbois account is problematic for a number of reasons:

A. The sale was made by November 5, according to Fr. Darbois' letter of that date. If Bishop Beaven's "threatening letter" had been written only days earlier, it would have dated from early November or late October at the earliest. However, in late October the bishop was apparently quite well-disposed toward the Assumptionists and expecting them to remain in Worcester indefinitely. On October 29 he wrote to Fr. Darbois, expressing approval of the establishment of a house for missionaries in Greendale, reserving for the future a decision on opening an alumnate or an agricultural or industrial college. (Beaven to Darbois, quoted in Moquin and Richards, *Assumptionists in the United States,* 46.)

B. According to a Darbois letter dated November 1, the bishop had already learned that the Assumptionists were about to purchase a house. (Darbois to Bailly, quoted in Moquin and Richards, *Assumptionists in the United States,* 46.) By that time there would not seem to have been any reason for the bishop to insist that they leave the rectory in the near future.

C. The community occupied the building on November 12. (Rochain, "Rapport sur la maison de Worcester [États Unis]," 1912 AG IH-7, 1), and the first mass was said there the following day. (Darbois to Bailly, November 13, 1903, AG 2OS 27, 1.) If November 12 was "only a few days after the threatening letter of Bishop Beaven," the bishop's threat would have occurred after the Assumptionists had made their purchase, an unlikely possibility and one that would contradict the claim that the ultimatum led to the purchase.

D. Fr. Darbois' 1925 account does not quote from the supposed Beaven letter. Neither do the subsequent accounts that repeat the story of the ultimatum. (For example, Odilon Dubois, A.A., "Le Collège de l'Assomption," *Seizième Congrès de l'Union Saint-Jean-Baptiste d'Amèrique* [n.p. 1947], 57; Polyeucte Guissard, A.A., "Un demi-siècle d'histoire, 1904–1954," *Assomption/Assumption, 1904–1954* [Worcester, 1954]: 146.) No such letter has been found in the course of the present inquiry.

67. The first Assumptionists to live in the Worcester house were Frs. Thomas Darbois, Isidore Gayraud, and Jean-Marie Laity, and two lay brothers, Bros. Philippe Uceda and Jean Despas.

68. Darbois to Bailly, Nov. 13, 1903, AG 2OS-18.

69. Fr. Darbois mistakenly recalled this event as occurring at the beginning of Lent in 1904. The date was corrected by Moquin and Richards, *Assumptionists in the United States,* 48.

70. Darbois, "Souvenirs de fondation," 59-60.

Chapter One

Alumnate into College (1903–1919)

As all Assumptionists knew, Emmanuel d'Alzon's 1844 vow to renounce "ecclesiastical dignities" had been taken at the shrine of Our Lady of Consolation in Turin, Italy. In Worcester, Fr. Thomas Darbois in 1903 noted that he had received Bishop Thomas M. Beaven's permission to establish the new Assumptionist house in Worcester around the feast of Our Lady of Consolation. Based on that coincidence, Fr. Darbois recommended the naming of the new institution "Our Lady of Consolation," and from Paris Superior General Emmanuel Bailly approved.[1]

It was one thing to give the institution a name, quite another to give it a clearly recognizable identity. It soon became evident — sooner to some than to others — that an alumnate on the French model could not be reproduced in Worcester. Also, as the effort to shape a successful alumnate was being carried on, some leading Assumptionists were also thinking about the other part of the educational enterprise once envisioned for Granby, an Assumptionist agricultural and commercial college. Would the creation of such a college in Worcester involve the transformation of the alumnate into a very different kind of vocational school? Or would the alumnate persist, while a separate, but somehow related, institution offered a Catholic education to young men interested in farming or business? Another course, the one the institution would actually follow, was hardly imagined during the first years of the Assumptionist enterprise in Worcester. By the end of World War I, the classical curriculum of the alumnate would become the foundation on which a training school for future priests was redefined as a classical college open to aspirants to the priesthood and to young Franco-Americans preparing for other professions as well.

Similarly, few if any Assumptionists in the early years of the Worcester foundation anticipated the dangers that would emerge from the simple fact that the Congregation spoke French. It seemed self-evident that they could put that aspect of their culture to good use in New England by ministering to French-speaking people, almost all of whom had come from Canada. The Assumptionists' French background seemed to be nothing but an asset, providing them with a natural link to a sizable and growing sector of the region's population. The early Assumptionists could hardly grasp the implications of binding themselves to a people whose clerical and lay leaders regarded cultural survival, *la survivance*, as a sacred cause. Little did they understand how they and their French-speaking school might be pulled into the rift that had opened between militant French-Canadian Catholics and the mostly Irish-American hierarchy of the Catholic Church in the United States.

As they wrestled with these complex issues involving the very identity of their Worcester institution, the Assumptionists would face the financial challenges that were to be expected by an organization seeking to open and expand a school. They would also face serious problems of leadership, some of which were both predictable and predicted.

Fr. Thomas Darbois in 1903 presented his superior general with a development plan for the Worcester house. The first step would be to pay off the debt of around $6,000,[2] after which the focus would be on raising enough money to construct an adequate building. With enough energetic activity, Fr. Darbois believed, the alumnate could be built and ready for students within six years.[3]

Because his plans, and those of the Congregation, were not limited to the alumnate, Fr. Darbois wanted Bishop Beaven to authorize the same activities he had allowed the Assumptionists to undertake in Granby: a missionary house with a chapel open to the public, an alumnate or apostolic school (the names were used interchangeably), and an agricultural and commercial college. Fr. Darbois had thus been annoyed by the bishop's October 1903 letter postponing consideration of educational endeavors while granting permission for a missionary house in Worcester. "When we purchased property here," Fr. Darbois assured Fr. Emmanuel Bailly, "I gave the information to the bishop and repeated to him, at least for the tenth time and in writing, that we were establishing ourselves at Greendale first to be missionaries, and once we had sufficient resources, to have an alumnate, and after that an industrial and commercial college."[4]

Fr. Isidore Gayraud, A.A.

On the leadership issue, Fr. Darbois' initial plan was straightforward. To get the new missionary house solidly established, he wanted to remain personally in Worcester for at least six months.[5] This would also be an opportunity to refurbish his reputation, because questions about his leadership had been raised after the Granby collapse. He did not want it said, "as alas was said about Granby, that I prepare difficult situations and that afterwards I leave it to others to clear things up."[6]

Fr. Darbois was less confident about finding long-term leadership for Worcester. Of the Assumptionists available to staff the Greendale house, the one with most seniority was Fr. Isidore Gayraud. However, some of the priests and brothers who might have to serve under Gayraud were hotly opposed to his selection. "The father, who is such a good religious, is excessively severe," Fr.

Darbois wrote to Fr. Bailly. "[H]armony would not reign in the community with such a head."[7]

Fr. Darbois' wish to remain in Worcester was granted, and during the winter of 1903–04 he and Frs. Isidore Gayraud, Marie-Jean Laity, and Georges Demiautte worked as missionaries in French-speaking parishes and also served as chaplains for communities of nuns and brothers.[8] The four priests managed to earn enough to meet the interest payments on the community's loans and to pay off a small part of the principal. The initial phase of Fr. Darbois' long-range plan for the alumnate seemed to be unfolding well. However, that plan, and Fr. Darbois' more pressing precautions about Fr. Isidore Gayraud, would not long prevail.

Superior General Fr. Emmanuel Bailly visited Worcester in the late spring of 1904. He was disturbed by the virtual absence of regular religious life among the priests and brothers who made up the community. The situation was an inevitable result of the fact that the priests had to be away from the house to do their missionary work. This strenuous effort was rewarding, both as a service to French-speaking Catholics and as a source of revenue, but Fr. Bailly worried that the itinerant priests might lose the spirit of community life. He was also concerned about the isolation of those who stayed behind, especially when the two lay brothers were left with no priest to exercise authority over them. The Assumptionists in Worcester, Fr. Bailly reasoned, needed a larger community, one in which there would be resident priests who would only rarely leave the house. That could be achieved by adding to the core of missionary priests a staff of priest professors.

"So it is urgent," he instructed his Worcester sons, "to establish the work that is supposed to be, along with the missions, the principal purpose of this foundation, that is, the alumnate. With it, a regular and permanent community can be completed, where the missionary will always regain his life of regularity, his life of prayer, his family life." Bishop Beaven, the superior general added, had told him formally that he would permit the opening of an alumnate, and in fact would be happy to see that happen. "Let us not delay in beginning," Fr. Bailly exhorted, "even if we have to do it on a very small scale and with only a few children."[9]

Having overridden Fr. Darbois' plan to spend six years preparing for the alumnate, Fr. Bailly before going back to Europe, also overruled the founder on the matter of leadership in Greendale; he asked Fr. Isidore Gayraud to take charge. Fr. Gayraud declared himself unsuited for the task, but as an act of religious obedience he reluctantly accepted responsibility.[10] Despite his selection by the superior general, Fr. Gayraud never received the official nomination papers that would have made him the canonical superior in Worcester.[11] While Fr. Gayraud bore immediate responsibility for local developments, he remained technically subject to the authority of Fr. Thomas Darbois, who would serve as common superior of the Worcester and New York houses.[12]

Fr. Emmanuel Bailly, A.A.

As Fr. Gayraud understood the arrangements, New York was to contribute $1,000 annually to help Worcester cope with its debts.[13] The superior general was to provide a loan — Fr. Gayraud believed the amount was to be $4,000 — to help the Worcester community purchase and refurbish a second, three-story building, located on a Fales Street lot adjacent to the house in which the missionaries were living. Before the opening of the alumnate, the two buildings were to be linked by an enclosed corridor about eighty feet long.[14] Several rooms on the first floor of the second building were to be combined into a chapel, leaving room on that floor only for a classroom or parlor; the second floor was to hold a study hall and two rooms for professors, while the third was to become the dormitory for the boys. The first building, with its tiny chapel closed, would house the kitchen, dining hall, and bedrooms for the missionaries and visitors.[15]

Montage published in 1920 showing: (top) 27 Fales St., first Assumptionist residence in Worcester; (bottom) adjacent building, linked to first by enclosed corridor, first alumnate building; (top right) Fr. M.-Louis Deydier, Superior, 1919–23; (bottom left) Fr. Omer Rochain, A.A., Superior, 1909–19; (center) sketch of Assumption College as it would look in 1922

Between the superior general's decision to launch the alumnate and the normal opening of schools in early September there were only three months to prepare. Halfway through that period, Fr. Gayraud in Worcester notified Fr. Darbois in New York that nothing had been done about the purchase of the second building, let alone the extensive rehabilitation required, including the installation of a heating system. He had yet to receive the $4,000 promised by Fr. Bailly, and now Fr. Darbois was refusing the $1,000 per year supposedly due from New York. Without that help, Fr. Gayraud wrote to Fr. Darbois, "we can do nothing, except to vegetate for an eternity; besides, you are the Superior; the debts fall back on you, and the bills will be sent for you to pay when there is nothing left in the bank."

Almost as urgent as the lack of money was the lack of staff. The priests already in Worcester were to continue to work as missionaries. Fr. Gayraud needed professors, and Superior General Bailly was supposed to be sending some. He had made contact with the Brothers of Christian Instruction of Ploërmel (Brittany) to supply an English teacher, since none of the available Assumptionists knew the language well enough to teach it. "Remind Fr. Emmanuel [Bailly]," Fr. Gayraud wrote to Fr. Darbois, "of the question of the brother of Ploërmel; we need him for Sept. 8; we could start with him, even if the [Assumptionist] religious didn't arrive until a month later."[16]

As the summer waned, there were some hopeful developments. With the help of a $1,500 loan from Fr. Bailly,[17] the second house was purchased on July 31.[18] On August 29, Bishop Beaven granted permission to open the alumnate's new chapel to the public.[19] Irish and French-speaking Catholics in the neighborhood lived so far from their parish churches that the cost of transportation was frequently beyond their means, so many were not attending Sunday services.[20] By saying mass and doing their best to preach in two languages in their own chapel, the Assumptionists would serve the neighborhood faithful and also realize some income, an arrangement that persisted until the opening of Our

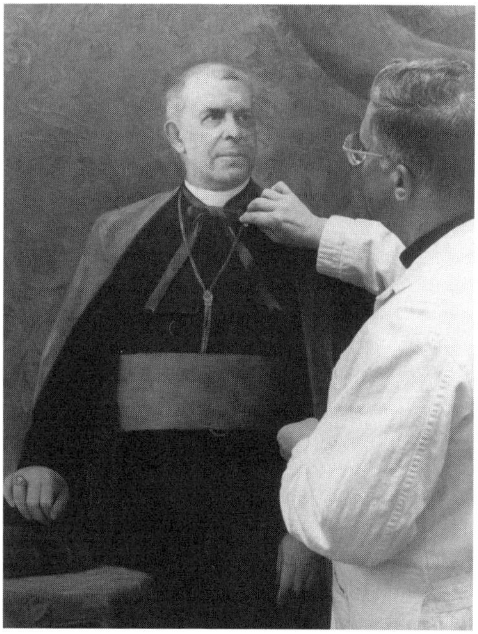

Portrait (1900) of Msgr. Elzear Brochu, being cleaned in 1990 by Fr. Donat Lamothe, A.A., College Archivist. Roger Trahan photo.

Lady of the Rosary parish in Greendale in 1911.

The projected opening date of September 8, 1904, arrived and passed, but for lack of a teacher the alumnate did not open. Disappointment arising from the continuing delay was to some extent eased by the encouraging news, received at the end of the month, that the last will of the recently deceased Monsignor Elzear Brochu included $10,000 for the Assumptionists of Greendale. Msgr. Brochu, pastor of Notre Dame parish in Southbridge, had been impressed by the Lenten preaching of Frs. Darbois and Demiautte in his church, and after conversations with them had promised a substantial bequest to support their educational work among French-Canadians.[21]

On the last day of September, Fr. Donat Cordier arrived in Worcester to serve as the first professor of the Alumnate of Our Lady of Consolation. Although the Ploërmel brother who was to teach English had not yet arrived (and would not until November 24), on October 1 the alumnate was officially opened.[22] Judging from Assumptionist practice in France and later in Worcester, the four original residents were probably between ten and fourteen years old. Fr. Cordier — who carefully instructed one correspondent that Greendale was pronounced "grinndèle"[23] — reported that he began with students of quite different abilities: "[T]wo know a little French even though they write it poorly, a third knows very little of it, the fourth, an Irishman, knows none at all."[24]

The Irish student, sent from New York, would soon depart.[25] The new professor was amused by odd expressions he heard in the language of the French-Canadians, but far less amused by the deficiencies in their academic preparation. "As for spelling," he wrote, "don't speak of it; they write French like a Mexican cow. You [might] think they know a little geography, a little of the history of their country, a little arithmetic. Alas!"[26]

As the lone professor, Fr. Cordier was also the only member of the religious community living in the house occupied by the boys, and he found it was "not possible to leave them for a minute."[27] He got the students out of bed, led morning prayer, supervised them during study periods, taught classes, and watched over them during recreation. Though he found his charges to be pouters, he was pleased that they came on their own to confess to him regularly and received communion afterward. "Their piety," he wrote, "without being profound, is real."[28]

By October 24, there were still four students in the alumnate, but not the same four. Three had left and had been replaced by new recruits.[29] The new arrivals were, Fr. Cordier wrote,"children who had no idea of what a college might be, one crying on my right and another on my left, needing a mother more than a professor."[30] In the course of the first year, the population fluctuated between four and nine. By June, the pioneering professor was still critical of his charges, but also still hopeful that they could be shaped in a positive way. "It's curious how there is something of the peasant in every Canadian, and something of the mule," he wrote. "All the same these children are not bad. But they are babyish and ignorant and careless and difficult! I have varieties of character, but I can say that a precocious pride and an astonishing vanity is at the core of each of them. Yet one can completely seize and capture this little society by its very vanity."[31]

Though Fr. Isidore Gayraud had managed to open the alumnate, he was not functioning effectively as the local superior. When Fr. Thomas Darbois visited from New York in late November, Fr. Gayraud took up his pen and wrote to the superior general, "Fr. Thomas has been here since yesterday, and he wishes absolutely that I write you to tell you that the house is going badly." Fr. Gayraud frankly acknowledged the truth of that observation,"for I know myself enough to know that it is impossible for me to be Superior. A Superior with authority and some mellowness could arrange everything; but those two qualities are absolutely missing in me."[32]

Fr. Thomas Darbois had come to Worcester the previous day to attend the organizational meeting of what would on December 1 become a new civil corporation in Massachusetts, the Society of the Augustinian Fathers of the Assumption. Fr. Darbois was elected its president, Fr. Gayraud was chosen treasurer and clerk, and, in addition to those two, the original board of directors was made up of Frs. Lazare Chaband, Georges Demiautte, and Donat Cordier, the first professor.[33]

On that same day, the long-awaited Brother of Ploërmel, identified in the sources only as Anthelme, arrived to become the alumnate's first English teacher.[34] The curriculum of that first year was probably shaped by Fr. Gayraud's judgment that a preparatory year would be required[35] before American students would be ready to begin, in the French language, the six-year classical curriculum typical of alumnates in France and of classical colleges in both France and Canada.[36] Compared to twentieth-century American practices, the first four years — after the preparatory year — would be roughly equivalent to a high-school or academy education, and the final two would be at the level of the first two years of college. In the European system, students who achieved the baccalaureate at the end of the six "college" years were eligible to go on to university. In a printed letter sent to French-Canadian pastors in 1905, Fr. Darbois announced that the "apostolic school" in Greendale taught Latin, Greek, French, English, the sciences, "and all that constitutes the program of a serious secondary establishment."[37]

Rather than provide a free education, as was the practice in the Europeans alumnates, the Assumptionists decided in Worcester to charge a fee of $10 per month for room, board, and tuition. This innovation, which some would later see as a critical deviation, was based on the advice of two French-Canadian pastors in Worcester who insisted that offering boys a free education "would only attract undesirables."[38] In Europe, alumnists were kept in their monasteries when school was not being conducted, in order to shield them from worldly attractions. In Worcester, a second divergence from European practice occurred at the end of July, when the students were allowed to return to their families for a six-week vacation.[39]

These modifications of the alumnate would soon become matters of controversy among the Assumptionists. However, in the fall of 1905, the two most urgent needs were to put up a larger building and to replace Isidore Gayraud as "superior." Counting on the promised Brochu legacy, Fr. Gayraud was addressing the first of these exigencies. He had already located a site not far from Fales Street and had submitted preliminary building plans to Superior General Bailly. The space problem that needed solving was fully evident when seventeen students crowded into the Fales Street house for the opening of the second school year.[40] The other urgent need was addressed at the end of October with the arrival of a newly appointed superior, Fr. Tranquille Pesse. Though he was canonically appointed superior in Worcester, Fr. Pesse would remain under the supervisory authority of Fr. Thomas Darbois in New York.

The new superior showed no more enthusiasm for leading the community in Greendale than his predecessor had. Tranquille Pesse had traveled peacefully from Istanbul to France to New York believing that in Worcester he would be what he wanted to be, a simple worker in the vineyard, ranked "the last in the house."[41] During his voyage, however, plans had been changed. Fr. Gunfrid Darbois, brother of the common superior in New York, had declined Superior General Bailly's appointment as superior in Worcester, using what was later described as the "pretext" that having two Darbois as superiors would invite the charge that America was becoming a "family fief."[42]

Two days after his arrival in Greendale as the new superior, Tranquille Pesse was barely recovered from what he described as the "melancholy into which I was plunged for a week" after hearing of his appointment.

"It was a great disappointment," he wrote to Superior General Bailly, "to learn of the new modifications, and I admit to you that it was very painful to submit myself to the new order of things. I sense that I am not up to the situation. I am not well-informed about anything at all; the responsibility terrifies me. All of these reasons suggested to me for a moment the intention of flatly refusing the heavy charge that you want so much to confer upon me. After mature reflection I made my sacrifice; whatever may come, I will do whatever I possibly can. If I do not succeed, so much the worse, but you will not be able to attribute the fact to anything other than my incapacity, of which I am well aware, and not to my bad will."

Turning to the other urgent business, Fr. Pesse asked for the superior general's prompt approval of the building plans submitted by Fr. Gayraud. Even if the first steps were taken immediately, he warned, it would be very difficult to have a new building ready for October of 1906. "While awaiting a response," he wrote, "I entrust the matter to St. Simon and St. Jude, patrons of desperate causes, whose feast we observe today."

"Next year," the new Superior concluded wistfully, "you can place me last in the house; it would make me the happiest man in the world."[43]

The plans the superior general was being asked to approve called for the alumnate to move approximately two hundred yards north from its location at the corner of Fales and Emerson Streets to a building to be erected on Baltimore Avenue (later Assumption Avenue), at the end of Emerson Street. Fr. Bailly, as was his custom, did not provide the prompt response the new Superior in Greendale was seeking. The lot intended for the new project was not purchased until February 26, 1906, and it was not until June 13 that a telegram from the superior general

Fr. Tranquille Pesse, A.A. with students and faculty, 1907

finally allowed a construction contract to be signed. As a result of the delays, when classes opened in September they opened for only twelve boys at the original site on Fales Street.[44]

One of those students later recalled that the Fales Street heating system was transferred to the new building long before the boys were able to follow. For lack of central heating, during a frigid fall and winter on Fales Street, kerosene heaters were carried from room to room as the boys moved from one activity to another.[45] Not until January 1907 were the veterans able to move into the new brick and granite building on Baltimore Avenue, designed to accommodate up to sixty-five or seventy students. They were joined by what was left of the preparatory class that had been scheduled to enter four months earlier.[46] In March, Fr. Pesse was able to report that the census would reach thirty-three after Easter.[47]

The priests in the alumnate tried, in the spirit of Emmanuel d'Alzon, to create a family atmosphere in the house. Fr. Donat Cordier, in the first weeks of Our Lady of Consolation, wrote of the fun of playing football during noontime recreation. Fr. Hydulphe Mathiot, who joined the house in October 1904 recalled a half-century later that the daily schedule, the religious exercises, and especially the spirit of Greendale matched those of the other alumnates. "Life was familial," he remembered, "to the great benefit of the students. . . . The fathers were continuously with the students, in study, at recreation." Isidore Gayraud, Fr. Mathiot recalled, was especially active in the games, and the students were "crazy" about him. "Hard and severe toward the fathers, he pardoned mischief when students were involved." It was Fr. Gayraud who took care of the boys' scrapes and bruises, mended their clothes, and when necessary sewed patches on their garments.[48]

Valmore Gaucher, who spent two years as a student at Fales Street, moved on to the new building, and later had a long career as a lay faculty member, fondly recalled outdoor games with the priests, picnic walks to distant Lake Qunsigamond led by Fr. Tranquille Pesse, biscuits handed out in class by Fr. Georges Demiautte to students who performed flawlessly on their recitations, and rehearsals under the demanding direction of Fr. Hydulphe Mathiot for a program of songs, recitations, and a short play presented by the boys at a downtown Worcester theater.[49]

While the Baltimore Avenue building was under construction, talk about transforming the alumnate into a French-Canadian college began to be heard publicly. Fr. J. Edmond Perreault, pastor of Holy Name of Jesus parish in Worcester, was particularly eager to see the Assumptionists educate a general population of French-speaking male students, not just those hoping to become priests.[50]

From their pastors French-Canadian parishioners heard that what was being built in Greendale was a college, similar to the colleges in Canada, though less expensive. French-speaking visitors flocked to the construction site on Sundays to observe the progress being made, telling one another, "This is a college for us."[51]

The superior, Tranquille Pesse, may have encouraged such speculation by considering Assumptionist sponsorship of a more general college, all the while referring to the institution he already led as a "college," rather than an alumnate or apostolic school. "If we had a commercial college," Fr. Pesse reported to Superior General Bailly, "it would be filled at the start of classes. Parents, despite the pressure of the majority of their pastors, are hardly enthusiastic about Canada and look very positively on the construction of the college." "[Even] if we could accept all the children who want to follow a classical course," he wrote on another occasion, "our new house would quickly be filled."[52]

As the nature of the institution began to become an open question, the school's bond with the French-Canadian population entangled it in two ongoing and related conflicts. On the one hand, French-Canadians determined to preserve their cultural heritage resisted — sometimes quite dramatically — efforts by Irish-American bishops to integrate them into an English-speaking church. On the other hand, French-Canadian "patriots" often had enough suspicion left over to direct a healthy dose toward members of their own group — or toward religious from France — who might be deemed insufficiently committed to the "Canadian cause."

Once Tranquille Pesse had moved his charges into the new building, he arranged a solemn dedication ceremony for January 28, 1907. The religious portion of the program was preceded by a luncheon to which the superior invited Bishop Beaven, "all the Canadian priests of the diocese and the most noted Irish priests." Bishop Beaven and the fifty priests and religious who gathered in the refectory heard a speaking program designed by Fr. Pesse to refute the rumor that the Assumptionists would, like other French

First wing of the college, constructed 1906–07

congregations, yield to pressure to abandon the French language and the Canadian population. He was also concerned about the charge made by Fr. Louis D. Grenier — pastor of Worcester's Notre Dame des Canadiens since the 1904 death of Fr. Joseph Brouillet — that the Assumptionists were not pursuing any consistent course with their school.

After the dedicatory banquet, one of the students recited in French a "compliment" to Bishop Beaven "in which," Fr. Pesse wrote, "the Canadian note was strongly accented." Toasts were offered to the pope, to the Canadian clergy, and to the French clergy.

When his turn arrived, Fr. Pesse set out to define in unmistakable terms the work of the alumnate: "Our house is open very specially to Canadian children who are directing themselves toward the priesthood; that is simultaneously telling you that we are following a consistent course. It tells you that we offer a classical education, as it is offered in similar colleges in Canada or elsewhere; and this classical education, whatever some people may say, whatever may be written in the newspapers, is given and will always be given in French, which is the language of the house. That is the very fixed, formal intention of all of our major Superiors."

This "profession of faith," Fr. Pesse reported to Fr. Bailly, "without giving offense to the Bishop, gave great joy to the Canadians."

Bishop Beaven, according to Fr. Pesse, "during the entire day showed an exceptional amiability toward us." When he rose to speak, the bishop, "ordinarily so reserved, responded in French to all the compliments that had been made to him; he declared that we were welcome, at home, in his diocese, to the great amazement of the Canadian priests, who could not believe their ears."[53]

Fr. Pesse's triumphal report on Bishop Beaven's reaction was less than thorough. Fr. Hydulphe Mathiot, who also witnessed the event, recalled years later that "the bishop was much less enthusiastic" than the Canadian priests about the tone of the meeting. As Fr. Mathiot remembered the bishop's response: "after having congratulated the fathers, especially on their work in his diocese, he insisted vigorously on the point that the religious should be universal, not attach themselves to one race, but serve all peoples, in imitation of the Church, their Mother."[54]

The bishop himself recorded his reaction. "I was not impressed," he wrote in his journal, "by the emphasis placed upon utterances announcing that the College was a French Canadian institution. It evidenced poor judgment." He found that the "statement — publicly uttered in an address read to the Bishop by one of the students... that the College was in the interest of French Canadian development was tantamount to open disrespect to the Ordinary [the bishop]. It was so considered by the majority of the clergymen present."

Bishop Beaven also found that "certain exaggerated statements" made to him in private by Fr. Thomas Darbois "savored of untruthfulness" and "placed the Community unfavorably in my opinions — I have been forced to fear that the Community has not come into the diocese for the General good."[55]

Bishop Beaven's suspicion that Fr. Darbois had been "untruthful" stemmed from a conversation between the two men just before the bishop left the Greendale campus. Bishop Beaven had asked Fr. Darbois whether the school the Assumptionists intended to develop was truly going to be an apostolic college, and Fr. Darbois had replied in the affirmative. Back in Springfield, however, the bishop noted in his journal, "This institution is named an 'Apostolic College' — where boys are received with a view of developing vocations for the Community — It is quite evident that all applicants will be received even with no intention of becoming religious."[56]

The bishop's "fear" was that the Assumptionists and their school might become a rallying point for Canadian "patriots" like Fr. J. Edmond Perreault of Holy Name in Worcester, who ardently championed the development of French-centered institutions inside the church and out. Indeed, the student who had recited the offending "compliment" to the bishop at the Greendale luncheon was a member of Fr. Perreault's parish, and at the conclusion of the boy's presentation, according to Fr. Mathiot, "the pastors applauded, [and] Fr. Perreault, a great patriot, wept with satisfaction."[57]

Fr. J. Edmond Perreault

Bishop Beaven's worries about the role the Assumptionists might play in the ethnic conflict did not suddenly arise during the dedication ceremonies in 1907. At least a year earlier he had come across a printed letter from Fr. Darbois, soliciting the support of the Canadian pastors of New England for the alumnate. Bishop Beaven privately found the letter "objectionable" because it was sent out without his permission and also because "[i]t announces an Institution essentially Canadian."[58]

Fr. Thomas Darbois had no doubt been sincere when he affirmed to the bishop that the Assumptionists were indeed conducting an apostolic school. The founder, however, had some suspicions of his own about the future character of the school. He was beginning to worry about Superior Tranquille Pesse's ideas about the development of a commercial college. In a letter to Superior General Bailly, Fr. Darbois took note of the Worcester school's deviations from the French model of the alumnate, and he expressed his worry that Fr. Pesse was intending to transform the alumnate itself into a commercial school.

In his deliberations about the school's future, Fr. Pesse had apparently put aside the "agricultural" part of the "agricultural and commercial college" the Assumptionists had envisioned for Granby and then for Worcester, but he was indeed very interested in starting a commercial college. He had in fact been in correspondence with Superior General Bailly on that very matter. Fr. Pesse saw the commercial college as an opportunity for the Assumptionists in the future, when the Congregation would have enough personnel to staff it. The program would correspond approximately, he wrote, "to modern teaching in the colleges of France; Latin is not studied, but there is more emphasis on sciences, accounting, etc." Demand among French-speaking parents in New England for such instruction was so great, Fr. Pesse believed, that it would take at least five or six schools to satisfy it.

The Assumptionists were being asked to open one in Worcester, and a friend had already offered "an important donation" to support it. The structure housing the school, the superior added, "would be entirely separate from the seminary." Fr. Pesse recognized that teachers were not yet available, but he wanted the superior general to understand that "it would be very desirable for us, and it would be vexatious if another congregation of French language were to come and plant itself beside us to do the very work that is rightfully ours."[59]

Two months later, in response to the "insinuations and fears" Fr. Darbois had expressed to the superior general, Fr. Pesse had to explain why he was no longer calling the school an "alumnate." In America, he wrote, that word suggested a meeting place for graduates of an institution, known as the alumni. Though he would have welcomed the opportunity to open a commercial college, Fr. Pesse flatly denied that he intended "to make of our alumnate a commercial college." On the other hand, he said, the alumnate had to be adapted to local circumstances, and he believed this was the real source of disagreement between him and the superior in New York. Fr. Darbois, said Pesse, insisted "that we treat our children here like our alumnists in France; that is impossible, unless one wishes to bury the work."

For example, Fr. Pesse repeated to Fr. Bailly, the sort of labor expected from poor alumnists in France could not be exacted from children now paying fees of more than $125 per year ($105 for the preparatory course). Similarly, American children could not be expected to go without a long summer vacation with their families. Neither they nor their parents would tolerate it.[60]

Fr. Darbois was not alone in his concern that the alumnate was losing its character. In the spring of 1907, Fr. Donat Cordier, the original professor, perceived a general lack of discipline in the institution. "The more I see," he wrote to a friend, "the more I perceive that our children are not at all alumnists. First of all they have the natural independence of the American, and one cannot manage them like alumnists. They pay their fees and they know it. . . . They are real collegians. To be American and to pay, that gives them an independence unknown in our modest alumnates."[61]

Fr. Pesse's juggling of the name of the institution continued to raise questions. In his semiannual report for June 1907, he left blank the section for alumnates provided on the Assumptionists' official form and instead submitted information in the section headed "Boys' Schools or Colleges," where he wrote in the word "Seminary." In 1908 he had to explain why he had changed the letterhead on his stationery from "Our Lady of Consolation, Apostolic School of the Assumptionist Fathers" to "Assumption College." That change, he assured Superior General Bailly had "in no way changed the nature or the purpose of the house." The school was generally known as "Assumption College," he said, and that was the name under which it was "inscribed in the books of the city."[62]

A new faculty member arriving in the fall of 1907 wrote to a friend, "I am professor at the seminary, at the college, at the alumnate (choose the name you want) of Greendale."[63]

Fr. Pesse continued to speculate about the possibilities of a "commercial" or "industrial" college, terms he apparently used interchangeably to designate a business school. In March 1908, he told Fr. Bailly that, considering the current debt carried by the Worcester house, "we cannot, without extraordinary assistance, have a commercial college," even though such a school was "craved" by all the Canadians of New England. "Someone offered us $37,000 to found an industrial college," he reported. "I replied that it was very difficult for us to undertake such a work, and also that such a college would require at least $100,000."[64]

In November 1908, Fr. Pesse even sought, without success, financial support for a commercial college from the republican government of France. He described the classical curriculum of "the French college of Worcester" without reference to its religious character. If resources permitted, he declared, "we would like to annex to the purely classical college a commercial college. We are being asked for it from all sides, and there certainly would be an abundance of students." He closed with the hope that France "would wish to accord us an annual subvention to help us pursue our eminently French goal."[65]

Fr. Pesse had good reason to be confident about the "abundance of students" available to populate both the school he had and the school he was contemplating. The new building constructed on Baltimore Avenue with a capacity of sixty-five to seventy students was soon filled. The 1907–08 school year began with fifty-five students.[66] A year later, sixty-six were admitted out of more than two hundred applicants, and ten were placed on a waiting list. However, as the school year opened, the superior in

Greendale was still in the dark regarding religious who were to be assigned to the school and the dates of their arrival, an operational problem that had become chronic and would remain so. On September 14, 1908, Fr. Pesse wrote to Fr. Bailly that he had still received no word about two replacements due for the opening of classes twelve days earlier. He had improvised in a number of ways, including taking on private instruction himself. Despite such difficulties Fr. Pesse could write, "We are already advantageously known on the outside, and our college has a good reputation."[67]

The problem of leadership was also a lingering one. Tranquille Pesse, who had accepted the office of superior with such great reluctance in October 1905, had repeated to the superior general as early as June 1906 that he was apprehensive about staying in that position. "I sense that I am not made for it, that I do not have a firm enough hand," he wrote. If he was to be kept on, "then I ask you please to give me a devoted subprior, well up-to-date on education, firm at the same time, who will occupy himself with the alumnate." During the next two years Fr. Pesse would often plead that he was overwhelmed with work as superior of the religious community, director of the alumnate, coordinator of missionary activities, treasurer of the house, substitute teacher, and substitute missionary. Not until the spring of 1908 was Fr. Omer Rochain designated to serve as subprior, the second in command.[68]

Overwork was a common problem in the house. In February 1907, Fr. Pesse again assured Bailly, "[I]f you want to discharge me completely, you would surely make me a happy man." Without a subprior or a treasurer, he had a house of sixty-five residents to manage; he was confessor to two communities of nuns; he heard the confessions of many of the boys as well; he had to teach eight classes per week. The professors were also overworked, with at least twelve classes weekly in addition to all the supervision required in a house with over fifty boys. They were complaining that they lacked time to prepare their classes and their Sunday sermons — necessary since many of the professors also went out to

Faculty in 1908. Front: Fr. Tranquille Pesse, A.A., superior; unidentified Ploërmel brother. Rear: Fr. Isidore Gayraud, A.A.; Fr. Georges Demiautte, A.A.; Bro. François Bourtembourg, A.A.; Fr. Donat Cordier, A.A.; Bro. John Despas, A.A.; Bro. Anthelme (Ploërmel)

parishes on weekends to earn an important share of the money that kept the institution alive. Under such conditions, they were losing their taste for teaching.[69]

In his own estimation and that of his colleagues, Tranquille Pesse was appropriately named; he was a man who truly yearned for tranquility and peace.[70] That trait led to severe criticism about his failure to exact discipline in his management of the school and in his management of the religious community. The criticism became especially painful to Fr. Pesse because Superior General Emmanuel Bailly encouraged his religious to keep him informed about how things were unfolding in their houses, and he had a tendency to believe the worst.

In 1907, Fr. Bailly heard reports of "great slackness" in the religious community, allegedly due to Fr. Pesse's failure to discipline wayward members. Fr. Pesse had in fact corresponded in great detail with the superior general concerning serious misbehavior by two Assumptionists who were eventually recalled to Europe. He was therefore disturbed at the tone of Fr. Bailly's demand early in 1908 that he explain his silence on several other matters that were now being reported to the superior general. "When I see that the religious make efforts to correct themselves," Pesse wrote, "that they carry out their duties faithfully enough, it is repugnant to me to speak of their faults in letters or reports that can at times fall into the hands of anyone at all." He preferred to encourage reform "by excess of goodness rather than excess of severity." Once again Pesse declared that he would be "delighted to leave my place.... I have reached the end, and I need a little calm."[71]

Divisions among the religious, complaints about the conduct of the alumnate, and speculation about its conversion into a college led Fr. Bailly to dispatch Fr. Ernest Baudouy in the spring of 1908 on a canonical visit in the name of the superior general. After talking with the ten members of the Greendale community, Fr. Baudouy urged strong leadership from the superior and due subordination among the religious. It was at that time that he appointed Fr. Omer Rochain to serve as Fr. Pesse's long-desired subprior.

In the light of what would occur a few years later, Fr. Baudouy's strictures regarding the school are somewhat surprising. He insisted that the Greendale institution was to be an alumnate and nothing else. "[L]et us not lose sight of the fact that this is a work of ecclesiastical vocations and that it should be organized according to that unique and exclusive purpose," he insisted. "The Very Reverend Father General did not ask for a college to be created here, but an alumnate." Granting that the European rules could not be directly transferred, it was the "spirit and customs" of the Europeans alumnates, not the practices of American schools, that had to be the guiding model. Though student vacations could not be eliminated, they were to be reduced as much as possible.[72]

Despite Fr. Baudouy's visit, tensions in the house persisted, broadening into open revolt. After seven more months, Fr. Pesse wrote to tell Superior General Bailly that, to all his previously expressed reasons for seeking a transfer, there was now added "the incessant malevolent criticisms about everything and sometimes nothing to which I am endlessly the object on the part of the religious of the house," several of them in particular. As superior he had done his "utmost to implant among the boys the spirit of family that reigns in the alumnates, and I can tell you that the boys love me, respect me, as well as the parents; the religious who criticize me find that I am too good, that I do not support them, etc.... [T]hey make a mountain out of nothing, insisting on exorbitant punishments, and wanted to turn the house into a veritable lycée[,] with the right to hit the boys thrown into the bargain."

Several had made it very clear "that they would be very happy and very pleased for the house if you removed me from my charge. You understand very well that under these circumstances I cannot stay where I am; it will be war all year, I will not be obeyed, I am not already. So for the good of the house, for the good of my soul above all, I entreat you to discharge me as soon as possible from the Greendale house."[73]

Three days later, the new subprior, perhaps reacting to the same incidents, also wrote to to the superior general to recommend that Fr. Pesse be transferred. Omer Rochain did not report open insubordination among the religious, but he agreed with an important part of Fr. Pesse's analysis of the situation. "Fr. Tranquille is a worthy man, wanting to please everyone, religious and boys, a friend of peace and of tranquility, suffering when he has to make an observation to the children or to punish them. He is loved by the children, because he has a good heart; that is not an evil."

"But," Fr. Rochain declared, "he is not an educator."[74]

Despite these conflicts, Fr. Pesse remained in place through the 1908–09 school year. In June, Fr. Ernest Baudouy made another canonical visit, reaching the less than surprising conclusion that "good understanding, cordiality, fraternal spirit, . . . and energy of government have left something to be desired." He once again urged "firm and bold" leadership from the superior and from the rest obedience, "easy, supple, without stiffness, without arrogance, without criticism, and without murmuring."[75]

The long leadership crisis was resolved when Fr. Omer Rochain rose from subprior to replace Tranquille Pesse as superior for the 1909–10 academic year. Serious interest in starting a business school seems to have departed with Fr. Pesse. Movement toward a classical college may have been foreshadowed in December 1909 when the Worcester founder, Fr. Thomas Darbois, was replaced as vicarial superior for New York and Worcester by Fr. Stéphane Chaboud.[76] According to Fr. Isidore Gayraud, who was still in Worcester in 1909, the "idea of transforming the alumnate into a college" had been suggested by Fr. J. Edmond Perreault of Holy Name parish during "the time of Fr. Tranquille, who liked the idea, but it was only toward 1908 or 1909 that Fr. Stéphane, superior general [*sic*], decided in a definitive way to dedicate us to the college."[77]

New thinking at the highest level was unmistakable when the superior general, Fr. Emmanuel Bailly, again visited Worcester in September 1910 for the third canonical visit in as many years. Fr. Bailly described the evangelization of "the Canadian populations" as the first purpose of the Worcester institution. "The second purpose," he said, "is the educational work called in succession alumnate, seminary, apostolic school, college." Whatever its name, he insisted, the establishment "should never cease to have as its unique intent the education of French-Canadians, but on the express triple condition of being a French-language establishment, of raising the children in the veneration of their nationality and of their traditions, and *of promoting among them priestly vocations.*"[78]

Only two years earlier, promoting vocations had been described by Fr. Baudouy as the "unique and exclusive purpose" of the alumnate. In Fr. Bailly's remarks it had become one of three distinctive qualities of a college whose "unique intent" was the general education of French-Canadians.

The decision to think of Assumption College not as an alumnate but as a classical college with a special interest in promoting priestly vocations may simply have been the recognition of a *fait accompli.* Bishop Beaven may not have been far from the mark when he concluded early in 1907 that it was "quite evident that all applicants will be received even with no intention of becoming religious." Tranquille Pesse in 1908 had concluded that only the most exceptional of the American children admitted to the school were real prospects for the priesthood. "[O]f the 45 children we now have," he calculated, "there are at least fifteen who are fairly serious and give some hope." Prospects for vocations might have been better, he believed, "if we could get them at a younger age, before they have been spoiled, but at ten years they barely know how to read and write. At 12 or 13 they are already perverted; they know everything, have seen everything, theaters not excepted." Parents removed their boys from the school "with the greatest of ease if their darlings find the regime too severe. Truly, this American education hardly favors religious or priestly vocations.... [T]hey have been raised too softly and too liberally and they love their comfort too much."[79]

When had the institution become more a college than an alumnate? Fr. Isidore Gayraud recalled that at the very beginning, at the 1904 inauguration of the Greendale school, "we sought above all, Fr. Emmanuel [Bailly], Fr. Thomas [Darbois] and I, priestly vocations." That changed, he said, when new Superiors took charge of the school. "Fr. Tranquille [Pesse] and Fr. Omer [Rochain] did not insist on the vocation." Perhaps the adaptation had begun even earlier. Fr. Hydulphe Mathiot, who arrived in Worcester in October 1904, traced the absence of rigorous screening for vocations all the way back to the opening that month on

Fales Street. Recruitment could not be carried on personally as in Europe, he recalled. Instead the opening of the school was announced in the newspapers, and "we took what came. We would find vocations somewhere in the pile." From the start, Fr. Mathiot recalled, parents seeking educational opportunities "placed their progeny with the fathers for purposes far removed from the priesthood." The first boys had had little commitment to becoming priests, he remembered, and those who entered after the new building opened in 1907 had no more settled vocations than the first.[80]

The alumnate evolved into a college on the Canadian model, Fr. Mathiot recalled, because it was with those Canadian institutions that Assumption College had to compete. The French-Canadian pastors described the school as a college to parents "who wanted to find at our place the same advantages" they would find at the Canadian colleges. When Omer Rochain became superior, Fr. Mathiot remembered, he went to Canada, studied the programs of Canadian colleges, and applied what he learned to his school in Worcester.[81] In 1911, indeed, the Assumption College catalog described the curriculum as "that of the ecclesiastical colleges of the Province of Quebec," with the addition of extensive study in the English language, "given the exigencies of the country."[82]

The change in the school's definition as expressed by Fr. Bailly in 1910 coincided with a major new emphasis on the association of the college with the French Canadians of the United States and Canada. In the past, the ethnic group had appeared in Assumptionist thinking mostly as objects of their apostolic services and as clients of their alumnate and college. Professors in Worcester had not infrequently expressed reservations about the backgrounds from which their students were drawn. A sense of cultural superiority had been sufficiently pronounced among some religious to draw from Fr. Baudouy in 1909 a warning against offending students by saying or doing anything that might suggest disdain for the Canadian or American nationalities or for the customs of those countries "when they contain nothing reprehensible, even if they are not always to our taste.... We are here not to make Frenchmen but to make good students and good priests."[83]

Fr. Bailly in 1910 depicted the French Canadians less as clients than as racial allies in a French Catholic struggle against Anglo-Saxon Protestantism. The common purpose of the college and of the missionary enterprise in Worcester, he said, was to help "the new or Canadian France, ... the French-Latin race distributed across America" to carry out "the mission given to it by God.... to expand the kingdom of Our Lord on the American continent and in the New World."

The Canadian people, Fr. Bailly observed, had had "the mission of introducing Catholicism" among "the Anglo-Saxon peoples" of North America and had successfully asserted their religious liberties "against the encroachments of a conquering England." The Assumptionists should envisage the extension of God's kingdom "to these Anglo-Saxon races of the New World" just as the Congregation was struggling "in the East [for] the return of the Eastern peoples or churches."

The Assumptionists' bond with the Canadians was to be made evident in the curriculum and the spirit of the college. "Let our professors... show esteem and sympathy toward all that Canada has done in the service of the Catholic cause, toward its heroes, its apostles, its apostolic origins and its meritorious struggles; let the national history of Canada be studied by our religious, let it be taught to the students; let its popular or patriotic songs be honored in the house; let no one ever amuse himself by mocking the Canadians or holding them up to ridicule."[84]

In this mood of racial and religious solidarity, during his 1910 visit Fr. Bailly presided over a celebration of the centennial anniversary of the birth of Emmanuel d'Alzon. The solemn High Mass was followed by a noontime banquet at which Worcester's "patriotic" Fr. J. Edmond Perreault rose and pledged $1,000 to help with the enlargement of the college's building.[85] Fr. Bailly

Fr. Omer Rochain, A.A.

himself offered a toast to "the *raison d'être* of the college." In words that would reverberate for decades, he proclaimed, "On the day the college ceases to be French, it should cease to exist, because it will no longer have its *raison d'être*."[86]

The superior general's vision of French-speaking Canadians spreading Catholicism across Protestant America seems not to have taken account of the fact that the majority of Catholics in the United States, including most of the nation's Catholic bishops, were of Irish descent and spoke English. Newspaper coverage of the ardently French and Canadian character of the d'Alzon celebration raised the ire of Bishop Thomas M. Beaven, something the superior, Fr. Omer Rochain, discovered when he visited Springfield some weeks later. Bishop Beaven declared himself "frightened by the spirit of the college." He had heard of the recent celebration at which the school was described as being "in favor of the development of the Canadian cause."

"Understand," the bishop said, "that there is only one cause, that of God. There is no Irish cause, Polish cause, Canadian cause. There is the cause of Jesus Christ. I am afraid of this college. I see it grow in this spirit with a certain terror." If the college continued to promote such a spirit, the bishop added, "in some time, instead of doing good, it will do evil, and a great evil." He warned that the Assumptionists, by taking sides with the fifty or sixty Canadian priests in the diocese, were alienating themselves from the other two hundred-fifty.

Alluding to Fr. Perreault's $1,000 pledge, the bishop spoke with agitation of "a small group of priests who want to be patrons of your work" and of people who were now proudly declaring, "We have our Canadian college." Bishop Beaven declared that he "did not authorize a purely Canadian college" and that members of other ethnic groups should be admitted to Assumption College if they applied.

The bishop had, of course, long known the ethnic character of the Assumptionist project. Fr. Rochain reminded him of the enterprise in Granby, where "it was well known that the work was to be Canadian," if only from Fr. Crevier's "spirit of chauvinism for the Canadian cause."

"This rivalry between Canadians and Irish," Fr. Rochain wrote to Fr. Stéphane Chaboud in New York, had not been so heated when Bishop Beaven first admitted the Assumptionists to his diocese in 1903. "For the last 2 or 3 years it's war, fierce war, and we appear, with our work, to be among the supporters of the revolt."[87]

Reporting to Superior General Emmanuel Bailly on his encounter, Fr. Rochain observed that the bishop seemed "bothered mostly by the publicity in the papers. The Irish must have criticized it. . . . There is no doubt that the bishop is very displeased. The war over the French and English languages continues, [and] is getting venomous. I do not know when it will end. The Bishop is afraid that with our influence we become the creators of a party, opposed to the Irish. . . . I think that he would like us to remain very quiet. That is what we will do. . . . For the moment, there is no trouble."[88]

Soothed perhaps by the Assumptionists' decision to "remain very quiet" on the ethnic front, Bishop Beaven's irritation apparently abated. There is no evidence that he tried to enforce his injunction that Assumption College admit boys from other ethnic groups.

There were already seventy French-Canadian students in the school, and so many applicants that Omer Rochain's focus during his first year as superior was on building an extension that would more than double the college's capacity. Ground was broken on October 15, 1910, for what was envisioned as the central section of a structure that would at some later time have a wing added to its northern end to balance the original 1906–07 "wing" at the southern end. Due in part to a month-long strike,

Central portion and tower, added in 1910–11 to original (1906–07) structure (right)

construction was still under way when 135 students reported for classes in September 1911.[89]

While borrowing for the building was adding substantially to the college's debt, Fr. Rochain was also buying up several parcels of land to extend and protect the campus. The college had been paying interest on its loans but not paying off the capital, so it still owed $30,000 on the 1906–07 building ($20,000 to a Worcester bank and $10,000 to Fr. Bailly and the Assumptionist Congregation). To build the new extension and to acquire the new land, it borrowed another $50,000 from a bank and an additional $40,000 from Fr. Bailly, bringing the consolidated debt to more than $120,000.[90] At the end of November 1911, Fr. Rochain observed that the debts of the college "appear enormous, and they could frighten those who do not know America. However, one must contract debts to arrive at anything."[91]

Servicing the debt would be a major preoccupation of Fr. Rochain during the eight remaining years of his superiorate. In 1911, he began asking French-Canadian priests each to take personal responsibility for paying a certain amount of the annual interest. He met with such encouraging success that he placed himself "on a collection circuit" to visit every Canadian pastor or vicar in every French-Canadian parish in New England.[92]

Fr. Rochain also worried about keeping enrollment at the level necessary to meet expenses. Earnings from missionary work, chaplaincies, and weekend parish work had from the start been the financial foundation of the institution. In 1910, however, for the first time fees from students made up more than half of the income.[93] Fr. Rochain's concern about that revenue source preciptated a debate around 1911[94] between the superior and his faculty over the practice of sending home boys who did not meet academic expectations. As Fr. Aurèle Odil recalled many years later, Fr. Rochain "asked us to keep the stragglers in the classes." The teachers argued that if there was a need for more income, a wiser course would be to raise the cost of tuition, room, and board. Despite his reservations, Fr. Rochain followed the faculty's advice, raising the cost from $120 to $160 per year. The strategy was not immediately successful; enrollment, down to 135 in September 1911 fell to 112 by May 1913. For whatever reasons, it was up again to 139 in November 1913, rose to 175 in 1915 and reached 180 — thirty beyond capacity — in 1917.[95]

With the leadership crisis in the past, with the institution defined and recognized as a Catholic classical college, with the issue of French identity under carefully managed control, Fr. Rochain set about consolidating and, in time, expanding the program.

The college continued to offer a six-year curriculum, preceded if necessary by a preparatory year at a reduced fee. In May 1913, there were twenty-six students taking the preparatory course, twenty-eight in the first year, twenty-five in the second, twelve in the third, five in the fourth, twelve in the fifth, and four in the sixth year.[96] Of the 112 students, only the last sixteen were studying at what was considered the college level in American education. The official policy was not to admit students until they were at least ten years old, but in 1915 students ranged in age from nine to twenty-two.[97]

Although the exclusive focus on priestly vocations had been superceded, the atmosphere of the college remained deeply religious. One priest recalled that religious life in Greendale was "intense, even compared to what I have seen in the best of our houses in France, colleges or alumnates." For students with "a serious priestly vocation," the League of the Sacred Heart was formed in 1912 in two sections, one for the younger students and one for the older.[98]

With some modifications, the daily schedule published in 1915 would prevail for many decades at the college. Students rose at 5:30; began prayer, meditation, and Mass at 5:55; prepared their morning lessons in the study hall from 6:45 until 7:40; and then had thirty-five minutes for breakfast and recreation. After morning class, which was conducted from 8:15 to 10:15, there was a fifteen-minute recreation period, followed by another ninety minutes in the study hall, during which students worked on written homework. Lunch and recreation filled the time between noon and 1:30.

The afternoon's work began with a forty-five-minute study period to prepare the afternoon lessons, followed by two hours of class ending at 4:15. After fifteen minutes of recreation students went to the study hall, where they worked on the afternoon homework assignment from 4:45 to 6:00. The period between 6:00 and 7:15 was taken up by supper and recreation. After that there was a final study period until 8:30, followed by prayers, and bed. During the students' fifteen-hour day, fewer than four hours were set aside for meals and recreation.

On Wednesdays and Saturdays, students followed the same morning schedule, but they were free from the end of lunch until 3:45, when they went to the dormitory to clean up. After that, the schedule was virtually the same as on full class days: a study period from 4:30 until 6:00, dinner and recreation until 7:15, and another study period until prayers at 8:30. On Sunday mornings, and on feast days, the students were allowed to sleep until 6:30. Religious exercises on such days occupied just over three hours of the day and evening, and study periods nearly five hours.[99]

The size of the religious community expanded rapidly during the first years of Fr. Rochain's superiorate. In 1910 there were twelve Assumptionists in the house in Greendale. At the end of 1911 there were sixteen. A year later, there were twenty-one. The first increase corresponded with the increase in enrollment that accompanied the opening of the new extension. The second had more complex origins.[100]

Like his predecessors, Fr. Rochain was sometimes less than satisfied with some of the priests and brothers working under his direction, and he regularly expressed his opinions about them to the superior general. Unlike his predecessors, Fr. Rochain experienced a welcome and dramatic change of personnel. Of the sixteen Assumptionists present in 1911, five were gone before the 1912–13 school year began. Not only were they all replaced, but the college received an additional infusion of five new men. When Bishop Beaven was asked to grant the newcomers permission to exercise their priestly faculties in his diocese, he termed the sudden influx "an invasion." Fr. Rochain pronounced the presence of the new men "a veritable treasure for the mission of Greendale" and regarded the absence of the departed as "a true blessing for the College."[101]

The change would prove to be a very mixed blessing, because there was a cloud over some of the transfers from Europe. The Assumptionists had not suddenly developed a host of new recruits eager to teach and preach in the United States. Superior General Emmanuel Bailly, dissatisfied with what he regarded as the liberal or modernist thinking prevailing in an important Assumptionist house of studies in Louvain, Belgium, had decided to dismiss the director of the house, Fr. Pierre-Fourier Mercklen, and to disperse many of the religious working there. Fr. Rochain in Worcester had been personally warned by Fr. Bailly to watch out for a special *esprit de corps*, a particular attachment to their old community, among the Louvainians being sent to Assumption College. At first the superior detected none. All of the new religious had "accepted without complaint" the work assigned to them, he reported in October 1912, and little was being said about the house in Louvain.[102]

By the spring of 1914, however, Fr. Rochain was describing three of his religious as "implacable Louvainians," all hungry for some communication with leaders from whom they had been separated, especially the dismissed director of the Louvain house of studies, Fr. Pierre-Fourier Mercklen. "I watch over the correspondence of these religious," Fr. Rochain reported to the superior general. "I doubt very much that the separation and the distance will be able to uproot this spirit of Louvain. Perhaps that will happen in the long run." In 1915, he was able to report that one of the unhappy Louvainians had undergone "an astonishing change," emerging from two years of melancholy to manifest "a constant and devoted attachment to the work of the college." [103]

By 1915, however, there was a much more astonishing change happening in the world. France was at war with Germany, and French law did not exempt members of the clergy from military service. French citizens, including those living at the college in Greendale, were being mobilized to serve their country, and the robust staffing level of the college had fallen victim to the Great War. In the spring of 1913, Fr. Rochain had presided over a community of thirty Assumptionists. By the end of 1915, he was down to fifteen.[104]

The war in Europe had broken out in August 1914, while Fr. Rochain and his staff were preparing for the start of a new school year. Mobilization orders quickly claimed two priests who had traveled to France during the summer as well as six priests and a lay brother still living in Greendale. Summoned to the colors himself, the superior felt a strong desire to leave Assumption and go home with the others. On August 10, he journeyed across Worcester to seek the counsel of Fr. J. Edmond Perreault at Holy Name. After a long talk with Assumption College's most ardent supporter among the diocesan clergy, Fr. Rochain walked into the Worcester post office and wrote a letter to his immediate superior in New York, Fr. Stéphane Chaboud. "My decision is made," he reported. "To save the work of the college, I will remain in Greendale. . . . My duty is to stay, and so I will stay."[105]

Fr. Rochain then traveled to New York, where the French consulate granted him a two-month reprieve to preside over the the reopening of classes, which began on September 8. In October, he went before a medical examining board and was placed in the French auxiliary service, with instructions to leave upon receipt of marching orders.[106] By the time those orders came in March 1916, jurisdiction had shifted to the ministry of war in Paris.[107] The most the consul in New York could do to help was to write a letter testifying to the importance of Assumption College as a French influence in the United States. Fr. Rochain promptly sent that evidence to Superior General Bailly, who was seeking help in the French capital.[108] In June, the Worcester superior was granted another three-month reprieve and told that he should continue seeking extensions through the consulate in New York.[109] For the rest of the war, a succession of petitions to the consulate or the ministry produced for the head of the college a succession of reprieves.[110]

During the extended emergency of World War I, Fr. Rochain was forced to rely more heavily than before on laymen as monitors to supervise student behavior. From the earliest years of the alumnate,[111] between one and four,[112] laymen had been engaged to teach each year. They had been assigned to English and science courses,[113] and some had even been entrusted with teaching the preparatory year.[114] The laymen were regarded as necessary but undesirable auxiliaries, because, unlike Assumptionists, they had to be paid, and their influence on the moral development of the students — or even of the religious — was suspect.[115] During World War I, there continued to be a few laymen teaching, but more were hired as monitors to help keep an eye on the students. By the summer of 1918, supervision of students was "almost entirely" in the hands of laymen.[116] "We have had recourse to laymen, given the necessity," Fr. Rochain remarked to the superior general, "but it is not the ideal, because their method of teaching, their spirit, their influence on the students sometimes leaves something to be desired."[117] As the war stretched on, he exclaimed, "How ardently I wish for the return of our mobilized professors, so we can get rid of the lay professors!!"[118]

Superiors in Greendale had also long complained about the difficulty and the expense of hiring servants, especially to do the cooking. Their preferred solution to that problem, too, was to be able to depend on religious rather than lay people. As early as 1905, Tranquille Pesse had explored the possibility of bringing in nuns, Oblates of the Assumption, "for the kitchen, the linen-room, and the chapel." When Fr. Rochain went to Quebec in 1911 to study the programs of Canadian colleges, he also approached several congregations of sisters, but could find none willing to commit themselves to Greendale.[119]

Only months after the outbreak of World War I, the long-sought solution presented itself in the form of a new religious community, the Sisters of Joan of Arc, founded by one of the Assumptionist missionaries in Greendale, Fr. Marie-Clement Staub. Although the new congregation would not be canonically established until 1920,[120] it began its community life when seven women moved to the Greendale campus on Christmas Day of 1914; they began their work at the college on the first of January 1915.[121] The sisters occupied a wooden building that twelve years earlier had been the original Assumptionist residence on Fales Street. In 1907,[122] the structure had been moved to Baltimore Avenue, which had been renamed Assumption Avenue by the time the sisters moved in.[123]

The mobilization of French Assumptionists had an immediate impact on the financial life of the college. Because all available personnel were needed in the classroom, the outside ministry carried on by the missionaries was suspended. Professors continued to go out to parishes on weekends and during vacations, but the revenue stream once provided by the full-time missionaries was sorely missed. Fr. Rochain periodically reported surprising acts of generosity on the part of the diocesan clergy, none more unexpected than $2,000 in cash from the seventy-seven-year-old proprietor of the Granby farm, Fr. Charles Crevier. After more than a decade of public hostility toward the Assumptionists, the Holyoke pastor promised also to remember the college in his will, understanding "that at his age he should not carry a grudge."[124]

As the cost of living rose dramatically in 1915, Fr. Rochain found it necessary to turn away students who could not pay their $160 room, board, and tuition costs in full. In the summer of 1917, he reported that the college had $7,000 in the bank and was meeting current expenses. It was not, however, amortizing its substantial debt. That year, he elected to add $80 to the annual student fee, and for the first time the charge was broken down into its component parts. The fact that students were charged $140 for their room and board but only $80 for their tuition testifies to the vital contribution the religious were making to the education of the students and to the financial life of the institution by working without pay.[125]

Although resources were sharply limited, Fr. Rochain and his collaborators during the war significantly expanded the offerings of the college and improved its standing among educational institutions. In 1916, Fr. Rochain inaugurated instruction at the level of the junior year of American college education and mapped out a curriculum for the senior year. He enlisted as the principal instructor of philosophy Fr. Florent Ancioux, once one of the most melancholy of the exiles from Louvain, but now fully committed to teaching in Worcester.[126] To support the feeble science curriculum, Fr. Rochain, with help from generous diocesan priests,[127] purchased what one of his professors later described as "a physics laboratory and a chemistry laboratory abandoned by another institution." Until then, science instruction had been given "without instruments and without a single demonstration."[128]

Fr. Rochain's goal was to find a way for the college to be able to award the bachelor's degree. Students finishing their six years of classical studies were well prepared to move on to seminaries or novitiates, but those not preparing for the priesthood often needed help getting into other colleges or universities to continue their studies.[129] According to Fr. Aurèle Odil, Fr. Rochain turned first to the original Augustinian Order, with whom the Augustinians of the Assumption were loosely affiliated as "third order" members. The Augustinians ran institutions such as Villanova College in Philadelphia that were authorized to grant degrees, so Fr. Rochain asked them to delegate that power to the college in Greendale. Failing in that quest, the superior hesitated to turn directly to the government of Massachusetts for a new charter, until Fr. Odil observed to him that "we were not risking anything. If [the state] refused we would lose nothing since we had nothing."[130]

At the end of 1916, the still anxious Fr. Rochain notified Fr. Emmanuel Bailly that he had "taken an important step, pregnant with consequences for the college, if it succeeds." It was because Assumption College was not officially recognized by the state that students were having difficulty getting into other colleges and universities, he explained. He had now requested that recognition. The process would involve an examination of the college's program by the state education commission, which would make a report to the "House of Deputies" [House of Representatives] and to the state Senate. "If the deputies and senators are kind enough, despite their bigotry and their anti-Catholic sentiments, to give their approval, a special law will be passed for us, a charter signed by the governor will be granted, and we will have our place and our rank among recognized colleges."[131]

Fr. Rochain's concerns evidently centered on the largely Protestant makeup of the state legislature and on the fact that the request was being submitted not only by a Catholic institution, but by one that did not have a single American citizen among its religious staff. The attempt to gain recognition might be audacious, he wrote, but "the longer we waited to do it, the less chance it would have, because the Protestants, noticing the ever-growing Catholic influence, will become less and less favorable in the granting of favors." To improve the odds, he had listed for the college's courses of Latin and Greek study "sundry pagan authors" as well as the Christian authors the students normally read.[132]

Fr. Rochain might also have been concerned by the fact that at Assumption College, only four students were working at the level of college juniors, and there were no seniors. Despite the superior's fears, the college's request moved smoothly through the Legislature. On February 6, 1917, with the signature of Gov. Samuel W. McCall, the corporate name "Augustinian Fathers of the Assumption" was changed to "Trustees of Assumption College," authorized by the Commonwealth of Massachusetts to award the bachelor's degree. Attributing the successful outcome to divine favor, obtained in part by the prayers of the students, Fr. Rochain granted them a three-day vacation to celebrate.[133]

According to Fr. Odil, Fr. Rochain's initiative before the state provoked a "very severe" response from his superior general, "as if he were prostituting the teaching of the congregation to the atheism of the state." Fr. Bailly had perhaps misinterpreted Fr. Rochain's sprinkling of "sundry pagan authors" into the curriculum submitted to the authorities. The superior general in Paris, Fr. Odil said to Fr. Rochain, would have to be told that "under the charter the religious retained the power to establish the curriculum and to award degrees following the program established by themselves." At Fr. Rochain's request, Fr. Odil typed out that explanation, which the superior signed and sent off to Fr. Bailly. "It shows," Fr. Odil sighed six decades later, "how difficult it was to make the true situation in the United States understood in Paris."[134]

In April 1917, Fr. Rochain could notify the superior general that Laval University in Quebec had sent its congratulations and its promise that "any student leaving Assumption College having obtained its diploma wishing to pursue his studies at the university will be admitted without an examination." Inspired by that response, Fr. Rochain proposed to seek similar recognition from the Catholic University of America in Washington, D.C., so that, in the catalog "we might print that the college, recognized by the state, has also been recognized by the two large Catholic universities of Canada and the United States. The impression would be excellent on the public as on the clergy." In March 1918, he could report, "On Feb. 16 Washington recognized our degrees. Thanks be to the Sacred Heart. I will now try at Harvard. That's the important point."[135] (No record has been found indicating any form of "recognition" of the college by Harvard University.)

In September 1917, as the college opened after three years of war, 180 students reported for classes.[136] They were to be taught by a faculty of thirteen, including three lay professors.[137] Although the proportion of faculty to students might have appeared healthy, the proportion of teachers to courses was another matter. The 180 students were distributed over nine grades, from preparatory to senior year in college. In the preparatory year, students studied five subjects. Beginning with the first year in high school, they studied at least seven.[138] Almost all professors were of necessity teaching different subjects and at several levels. Fr. Rogatien Bahuaut, for example, was professor of Latin, Greek, French, and history; Fr. Rumold Spinnaël taught geometry, physics, astronomy, and geology; Fr. Aurèle Odil was professor of apologetics, history of philosophy, Latin, and French.[139]

In 1918–19, Fr. Odil was responsible for one of two sections of high-school freshmen. He taught those students catechism, French, Latin, Greek, and history, while they studied English, algebra, and natural sciences under lay professors. Fr. Odil also taught apologetics to college juniors and seniors, as well as Latin, French, and history of philosophy to college seniors.[140] Under these circumstances, Fr. Rochain observed, some courses in catechism, history, and the sciences were inevitably neglected, and only with the greatest difficulty could they be completed according to the published program.[141]

When the United States entered World War I on April 6, 1917,

French officers visiting the college during World War I

the college's ties to France took on a new significance. Official missions sent by the French government to "maintain and expand the French influence" made a point of calling at Assumption. French officers on special mission gave Fr. Rochain advice on how to "use influential people in Paris" to have the mobilized Assumptionists sent back to Greendale as a means of "spreading the French influence," advice he promptly passed on to Fr. Bailly in Paris.[142]

With the college identified more forcefully than ever with France, and with the authority in hand to award bachelor's degrees, Fr. Rochain set out to cultivate in a new way the financial support of the French-Canadians of New England. In February 1918, the superior took some of his advanced college students to Fitchburg, Massachusetts, for the organizing meeting of the

Franco-American Catholic Federation. He returned with a promise that the new organization would raise $50,000 for the college by the end of the year. The promise had been reinforced by more than $2,000 in pledges from the delegates, to be paid by July 1.[143]

Organized by Joseph Lussier, editor-in-chief of *La Justice* in Holyoke, Massachusetts, the campaign focused on the college's need for additional building space. With 180 students, the institution had clearly outgrown a building designed for 150. Assumption College was depicted in the campaign as an indispensable resource for a people under cultural siege. "In this century of struggle," the federation said, "the victories of the future will be won by those elements who have understood how to produce within their ranks an elite of educated men, able to shine in the priesthood, in the liberal professions, or in the higher spheres of industry or politics." For that reason, the federation was pledging its support to "a house that Providence seems to have raised up in the heart of New England for the salvation of our people, Assumption College in Worcester."[144]

In a week of pledging in June, the federation surpassed its goal. By October, Fr. Rochain was reporting a total of $60,000 in pledges.[145] In the margin of the logbook used to record significant events in the life of the Assumptionist community in Worcester, Fr. Rochain drew a hand pointing to his admonition, "*May the Superiors of the college always remember that the college must to the last day of its existence remain a Franco-American college* to exercise its influence over the Franco-Americans of New England." Citing the earlier pledges of Frs. Pesse and Bailly, he added, "The promise has been repeated several times."[146] Despite the original goal of the drive, most of the proceeds were used to buy land to expand the campus.[147]

The federation's ideal of "an elite of educated men" reflected the official ideology of Assumption College. The 1917–18 college catalog, after emphasizing vocations to the priesthood, added, "while it is very necessary for the Church to have zealous and educated priests, it is no less important for it to find, among the faithful, those elite souls who, as a result of their thoroughly Christian training and instruction, exercise a salutary influence in the different professions they choose."

The classical curriculum's prescriptions for intellectual development were similarly premised on the need to form exceptionally skillful thinkers, writers, and speakers. The emphasis was unmistakably on literary studies, as was evident from the catalog's overview of the high school curriculum:

"To develop in the young man serious intellectual habits — precise thinking, neatness of style, discrimination, and a taste for nuance — one must first of all make his spirit supple through exercises of analysis and synthesis, make him understand his language in its smallest nuances, expressed by a different mood, a simple turn of phrase, the preferred use of a preposition, etc.

"Thus one must procure for the child a profound, complete, and sure knowledge of the rules of grammar of each classical language. The explication of the Latin, Greek, French, and English authors must, in these grades, extend even to the details, to enrich the young intelligences with ideas appropriate to their age and to make them skillful in finding clear and agreeable expressions of them." To this the catalog added curtly, "In mathematics and in other subjects, the curriculum reproduces the complete high school program of the state."

The college-level curriculum, with the new physics and chemistry laboratories installed, covered four years for the first time in 1917–18. The finished curriculum, the catalog claimed, "completes and crowns the study of the languages and of their literature, of mathematics and of the physical and natural sciences. One can readily understand that the education of a young man remains incomplete as long as he has not received this formation of the intelligence, which is the work of philosophy and of that ensemble of studies called by the ancients *humaniores litteræ.* This expression indicates perfectly the goal of the classical curriculum: to form the man in the child, by developing the intelligence, the faculty that is slowest to ripen and, nevertheless, the queen of all the others. Without this general culture, all specialization becomes premature.

"The studies prescribed in this curriculum are regarded as essential and necessary in every liberal education. Consequently, nothing is left up to the choice of the students."[148]

To promote an intellectual elite within the student body, Fr. Rochain as early as the fall of 1912 had created two literary circles, one for the younger students and one for the older, made up of "the best students in all the classes." The members met regularly, prepared for public declamations, and produced a printed monthly literary journal, *Vers l'Idéal* [Toward the Ideal]. True to the vision of raising a specifically French-Canadian elite, the members of the literary circle were encouraged to think and write about French Canada as their country and to regard themselves as the cultural heirs of the founders of New France. In a letter saluting the new publication, Superior General Emmanuel Bailly congratulated the literary circles on their "Catholic and Canadian hearts." He encouraged them to pursue their high intellectual goals "not only as intellectual young men, but also as the sons of those valiant Canadians who founded your noble country with their faith and their blood." He urged the students to be "above all Catholic and supernatural like them!"[149]

The college passed a milestone in its history on June 14, 1918, when the annual distribution of prizes was highlighted for the first time by the conferral of bachelor of arts degrees. The magna cum laude recipients were Anatole Desmarais and Romuald Crispo, both from Fall River, Massachusetts.[150] Desmarais became a priest, and Crispo a lawyer,[151] not unlike many other students who, as in past years, were "finishing" though not graduating from Assumption. In all, fourteen students "finished" in 1919, seven after completing the old six-year curriculum — or halfway through college — and seven from the junior and senior classes. Of the fourteen, nine were bound for the seminary, while the others planned to pursue medical or legal studies. "How many large colleges," Fr. Pierre-Celestin Régnier remarked, "would be proud to be able to present. . . nine future priests."[152]

Fr. Albert Robillard, the first man to be ordained after studying at Assumption, had celebrated mass in the college chapel in June of 1915. However, while the institution was producing a steady stream of priestly vocations, it was not producing new Assumptionists. Fr. Donat Cordier had pointed out as early as 1905 the need for a novitiate in America, because students simply could not be expected to cross the Atlantic to continue their studies and religious formation. In 1914, Fr. Rochain had made the same observation, noting that two students interested in religious life had entered the Marist novitiate in Washington rather than travel to Europe to join the Assumptionists.[153]

The first bachelor's degrees were awarded only three days before the Franco-American Catholic Federation launched its successful week of pledges for the college. Although Fr. Rochain and his staff could take some satisfaction from these wartime accomplishments, the desperate shortage of religious personnel not only persisted but got worse.

During the mobilization crisis of 1914, Fr. Rochain had fought off the notion that the college would not be able to open. "If we had closed the college this year," he wrote two months later, "it would have been a mortal blow for us." In 1918, he addressed himself to Fr. Joseph Maubon, who, as vicar general, was presiding over the Assumptionists following the death of Superior General Emmanuel Bailly on Nov. 23, 1917. In the middle of the June pledge week, Fr. Rochain sent Fr. Maubon a cablegram: "Absolute necessity to send four professors before the end of August. [Fr.] Vital [Chaffart] mobilized [and] returns [to France]." In a letter following up his cable, the superior explained the perilous state of the college's staffing, now that Fr. Chaffart had been called into service. "If reinforcement cannot be sent to us from Europe," he wrote, "I will find myself in the necessity of having to close two or three classes. Such an act would be disastrous for the college, at the moment when people on all sides are paying attention to us."[154]

In late August 1918, Fr. Rochain learned that several of his lay professors had also been swept up, this time in the American mobilization that followed United States entry into the war. On September 8, 1918, Fr. Pierre-Celestin Regnier, prefect of discipline and second in command in Worcester, wrote the vicar general that, even though classes would open in eight days, and even though more than one hundred-eighty students were expected, and even though several large rooms and the infirmary were being converted into dormitories, no word had yet been received about the arrival of teachers promised from Europe.[155]

"I will always remember the difficulties of the opening of classes in 1918," Fr. Rochain wrote in October. Despite the "penury of professors," more than two hundred students had been admitted. The superior had been forced to select several students "to help us as professors." One of the student-teachers was Louis Robert, who in 1928 would become the first Assumptionist priest born in the United States. In 1918, though only a college freshman, he was teaching geometry to high-school sophomores.[156]

The unprecedented number of students enrolled may have been related to the fact that fifty-two of them were having their tuition, room, and board subsidized by the government of the United States. Students subject to conscription were in uniform, members of the Students' Army Training Corps. To house them, a building had been bought and moved onto college land, where it was refurbished at government expense. Before morning classes, the enrollees drilled under the command of two young lieutenants. Sometime during their crowded academic day, they took military courses, after which they drilled again. In the evening, the student-soldiers were conducted back to their dormitory by Fr. Aurèle Odil, who spent the night there as their college monitor. Formed just one month before the Armistice in Europe, the unit was dissolved on December 4, 1918.[157]

Although the cessation of hostilities provided little immediate relief for the college, it moved Omer Rochain to reflect on what had happened and to project his vision for Assumption College into the future. "We have been able to cope with difficulties through an excess of work and through different contrivances," he wrote eighteen days after the Armistice. "But this state of affairs must disappear at the very earliest for the good of the college"[158]

"The Sacred Heart, it is my firm conviction, has blessed our college in a visible manner," he wrote in January 1919. "Despite the war it has prospered, philosophy [the junior and senior years of college] has been established, academic degrees have been obtained, confidence in it has grown, the fund drive in its favor is succeeding, its reputation [is] extending into all of the neighboring states. God has used our weakness to accomplish great things. This work must now be fortified. We have no right to betray those who have hope for its beneficent future for the Franco-Americans."[159]

In that same letter, Fr. Rochain asked permission to return to France himself in June 1919, to visit his aged parents, after an absence of seven years "and no doubt for the last time." He hoped also to use the trip to lay out for the vicar general the "true situation" in Worcester with respect to the religious community, the college's finances, and its future needs. Permission for the trip was granted, and by mid-April Fr. Rochain had arranged to leave from New York at the end of May. His journey, however, had taken on a new purpose. He had been exposed to disturbing evidence indicating that significant individuals and groups in Worcester had lost confidence in his leadership. "If you could find a way to replace me as Superior in Worcester," he wrote to Fr. Maubon," it would be a marked favor for which I would be very grateful to you, because sometimes the burden of responsibility is very heavy. We will talk about it soon."[160]

By July, writing from his parents' city of St. Omer, Fr. Rochain sounded even more discouraged. "I no longer feel in myself the strength to continue to carry such heavy responsibilities," he wrote. "I have already spent 13 years in Greendale, 10 as Superior. I've done my part. I am convinced that a renewal is necessary and even urgent. No one could ever imagine all that I have suffered at Worcester, especially during the war."[161]

Fr. Rochain had long regretted what he saw as the personal failings of some of the religious it had been his duty to live with, work with, and try to lead. "I try to maintain the supernatural spirit in the community," he had written in 1917, "as well as a great charity. That's the difficult point: charity. There are so many thin-skinned and egoistic characters! The office of Superior is not an enviable one. In religious life there is no lack of opportunities to practice patience." Now in France, Fr. Rochain contemplated a letter he had just received from Hermés Fuchs, a brother preparing for the priesthood who had spent the war years in Worcester. Fr. Rochain came to the conclusion that "certain religious, alas... have made efforts to denigrate their Superior, inside and outside, instead of supporting him." Under such circumstances, he wrote,"the thought of going back to Worcester saddens me profoundly."[162]

Brother Fuchs, who was also in France, told Fr. Rochain he should not return to Worcester except to pack his bags.[163] The brother then addressed himself directly to the vicar general and presented a stark case for Fr. Rochain's replacement. "Opinion at the college and on the outside is no longer favorable to Father Omer," he wrote. "His return to the college as Superior would be a great disappointment for all. Father Omer has done a great deal for the development of the work, but it appears that he is worn out, that he is no longer equal to the task. (That's the expression often used.) People are tired of him. He has lost almost all authority among the religious and among the older students."

Brother Fuchs granted that some of the professors might have had something to do with lowering the student's esteem for the superior, but he observed that Fr. Rochain had failed repeatedly to respond in a timely way to public demonstrations of disrespect toward him. He reported discontent among the federation that had just raised $65,000 for the college, including threats to launch a campaign against the school "if the Canadian cause were not promoted more." Among alumni, too, he reported dissatisfaction with "the attitude of the college."[164]

Despite Fr. Rochain's constant efforts to preserve and promote the French-Canadian character of the college — or, to use the more popular expression of 1919, the Franco-American character — he was being harshly criticized for what he regarded as one of his most significant accomplishments, securing from the state the authority to award degrees. The superior had pursued that course in order to help his graduates move on to further studies in American universities, and that, in the view of some Franco-American leaders, was precisely the problem. Fr. Rochain had opened a tempting door to Americanization. Franco-American young men would find it easier to enroll in American universities, a course that would put at risk their faith, their language, and their attachment to French-Canadian tradition. In becoming "Americanized," they would become defectors from their own people.[165]

"I have no ambition for the burden," wrote Omer Rochain, "and if it is judged appropriate to remove it from me, I will sing a *Te Deum* with all my heart."[166] At the end of August 1919, the forty-six-year-old superior was granted his release.

Notes

ABBREVIATIONS
AC — Archives of Assumption College, Worcester, MA
AG — Archives of the Assumptionist General House, Rome
AP — Archives of the Assumptionist Provincial House, Brighton, MA
PA — Archives of the Assumptionist Province of Paris
SP — Archives of the Diocese of Springfield, MA

1. Richard A. Richards, A.A., *The Assumptionists* (New York, 1980), 8; Darbois to Bailly, Nov. 13, 1903, AG 2OS-21. I am grateful to Fr. Wilfrid J. Dufault, A.A. for identifying the feast of Our Lady of Consolation as one specially observed by Augustinians on September 4 of each year, though it is not found in the regular liturgical calendar of the Catholic Church.
2. Darbois to Bailly, Nov. 13, 1903, AG 2OS-21.
3. Darbois to Bailly, Nov. 13, 1903, AG 2OS-21; Sept. 25, 1903. AG 2OS-19.
4. Darbois to Bailly, Nov. 27, 1903, Rome Archive 2OS-20.

   On this document the name of the month is typed over and could be read as "Sep." or as "Nov." An archivist has by hand dated the document "27 IX 1903" (Sept. 27, 1903). However, the letter states that as of its writing the Assumptionists were "completely installed in our new house." The Assumptionists did not move into their new house until November 12.

   Just before purchasing the Fales Street residence, Fr. Darbois had, as he claimed, given Bishop Beaven written notice of the Assumptionists' plans and goals, inviting the bishop to express any opposition he might have to the course the Congregation was laying out. See Darbois to Beaven, Oct. 27, 1903, bound letters volume [1903-7], 229-31, SP.
5. Darbois to Bailly, Sept. 22, 1903, AG 2OS-18.
6. Darbois to Bailly, Nov. 13, 1903, AG 2OS-21.
7. Darbois to Bailly, Sept. 22, 1903, AG 2OS-18.
8. "Archives de la maison," I, 2, AP.
9. Emmanuel Bailly, "Carte de la Visite Laisée à Notre-Dame de Consolation, Greendale, Worcester," June 3, 1904, AG 2PK-14.
10. Gayraud to Bailly, Jan. 18, 1908, AG 2OS-122.
11. Isidore Gayraud to Odilon Dubois, Oct. 12, 1946, AC.
12. Isidore Gayraud to Wilfrid J. Dufault, April 18, 1947, AC.
13. Gayraud to Darbois, July 13, 1904, AG 2OS-27.
14. Before construction the distance was estimated as 15 meters (Gayraud,

"Maison de Greendale," July 3, 1904, AG 2PK-13). Valmore Gaucher, who lived in the alumnate as a student, remembered the corridor as being eighty feet long. ("Souvenirs d'un Ancien," *Almanach de l'Assomption* [Dec. 1925], 18.)

15. Gayraud, "Maison de Greendale," July 3, 1904, AG 2PK-13.
16. Gayraud to Darbois, July 13, 1904, AG 2OS-27.
17. Darbois, "Etat financier de la maison de Worcester (Mass.)," April 8, 1905, AG 2PK-17.
18. Moquin and Richards, *Assumptionists in the United States,* 48.
19. Clodoald Sérieix, "College de l'Assomption, Worcester, Mass., Notice sur sa fondation et sa situation actuelle," 1928, 2, AP.
20. Gayraud, "Situation de Greendale," July 3, 1904, AG 2PK-13.
21. "Archives de la maison," I, 3-4 AP. The bequests made by Msgr. Brochu exceeded the assets of his estate. Once adjustments had been made, the Assumptionists in the second half of 1906 received a total of $8,500. (Tranquille Pesse to Emmanuel Bailly, Dec. 14, 1905, AG 2OS-47; Dec. 10, 1906, AG 2OS-70.)
22. "Archives de la maison," I, 4, AP.
23. Cordier to "un Religieux de Rome," June 21, 1905, AG 2OS-44.
24. Cordier to Emmanuel Bailly, Oct. 8, 1903, AG 2OS-30.
25. Anon., "Quelques notes sur la fondation de Greendale," mss appendix to Ernest Baudouy, "Carte de la Visite," April 27,1908, 25, AG 2PK-19; Lazare Chabant to Bailly, Oct. 27, 1904, AG 2OS-29.
26. Cordier to Ernest Baudouy, Nov. 16, 1904, AG 2OS-35.
27. Cordier to Bailly, Oct. 8, 1903, AG 2OS-30.
28. Cordier to Ernest Baudouy, Nov. 16, 1904, AG 2OS-35.
29. Lazare Chabant to Bailly, Oct. 27, 1904, AG 2OS-29.
30. Cordier to Bailly, June n.d., 1906, AG 2OS-74.
31. Cordier to "un Religieux de Rome," June 21, 1905, AG 2OS-44.
32. Gayraud to Bailly, Nov. 24, 1904, AG 2OS-26.
33. Minutes of the Corporation, 1-2, AC.
34. "Archives de la maison," I, 4, AP.
35. Gayraud to Darbois, July 13, 1904, AG 2OS-27. When a prospectus was published in 1906, the first year of instruction carried a lower tuition cost and was explicitly designated as preparatory.
36. Emmanuel d'Alzon's plan for a five-year version of this education was in time abandoned for the more traditional six-year course. (Polyeucte

Guissard, A.A., *Un siècle d'histoire Assomptioniste* [Worcester, 1950], 72.)

37. Darbois to "Monsieur le Curé," n.m, n.d, 1905, manila folders, SP.
38. Mathiot, Hydulphe, A.A., "Histoire Ancienne," typescript, 7, AC and AG 2PN-49. A handwritten note on the cover of the copy in the Archives of the General House states that the work was written in Walden, N.Y. in 1956.
39. "Archives de la maison," I, 5, AP.
40. "Archives de la maison," I, 5, AP. A 1908 account states that sixteen students reported in September 1905. (Anonymous, "Quelques notes sur la fondation de Greendale," mss appendix to Ernest Baudouy, "Carte de la Visite," April 27, 1908, AG 2PK-19.) In this and in many other instances, the precise number of students is difficult to establish, due to conflicts in the sources, which are probably due to students arriving and leaving during the course of the year. Subsequent statements about the size of the student body will be based on the author's best judgment, based on all available sources and should not be taken as exact measurements.
41. Tranquille Pesse to Emmanuel Bailly, Oct. 28, 1905, AG 2OS-46.
42. Anon., "Quelques notes sur la fondation de Greendale," mss appendix to Ernest Baudouy, "Carte de la Visite," April 27, 1908, 27, AG 2PK-19.
43. Pesse to Bailly, Oct. 28, 1905, AG 2OS-46.
44. Moquin and Richards, *Assumptionists in the United States,* 50; Pesse to Bailly, June 16, 1906, AG 2OS-67; Pesse to Bailly, Sept. 30, 1906, AG 2OS-69.
45. Valmore Gaucher, "Souvenirs d'un Ancien," *Almanach de l'Assomption* [Dec. 1925] 21.
46. "Archives de la maison," I, 7, AP.
47. Pesse to Bailly, March 30, 1907, AG 2OS-82.
48. Cordier to Bailly, Oct. 8, 1904, AG 2OS-30; "Archives de la maison," I, 4. AP; Mathiot, "Histoire Ancienne," 10-11, AC.
49. Gaucher, "Souvenirs d'un Ancien," 19-22.
50. Isidore Gayraud to Odilon Dubois, Oct. 12, 1946, AC.
51. Mathiot to Odilon Dubois, Feb. 19 [?], 1946, AC.
52. Pesse to Bailly, Aug. 1, 1906, AG 2OS-68; Sept. 30, 1906, AG 2OS-69.
53. Pesse to Bailly, Feb. 3, 1907, AG 2OS-80.
54. Mathiot, "Histoire Ancienne," 14.
55. Beaven journal entry, Jan. 28, 1907, "Records, Diocese of Springfield, 1871 to 1909," bound volume, 690, SP.
56. Mathiot, "Histoire Ancienne," 14; Beaven journal entry, Jan. 28, 1907, "Records, Diocese of Springfield, 1871 to 1909," bound volume, 690 SP.

Although the Assumptionists wanted very much to recruit new members for their own religious community, Bishop Beaven was inaccurate in stating that they were seeking only Assumptionist vocations. As in the European alumnates, students completing their studies in Worcester would be free to choose the diocesan priesthood, join the the Assumptionists, or seek membership in another religious community.

57. Mathiot, "Histoire Ancienne," 13-14.

58. Beaven note on the reverse side of Darbois to "Monsieur le Curé," n.m., n.d., 1905. manila folders, SP.

59. Pesse to Bailly, Jan. 10, 1907, AG 2OS-79.

60. Pesse to Bailly, March 30, 1907, AG 2OS-82.

61. Cordier to "Frère Agapit," June 8, 1907, AG 2OS-92.

62. Pesse, "Rapport semestriel," June 27, 1907, AG 2PN-65; Pesse to Bailly, Nov. 22, 1908, AG 2OS-111. According to an account published in 1946, it was Fr. J. Edmond Perreault, pastor of Holy Name of Jesus in Worcester, who suggested the definitive name "Collège de l'Assomption." (Odilon Dubois, "Le Collège de l'Assomption," *Seizième Congrès de l'Union Saint-Jean-Baptiste*, n.p., 1946, 59.)

63. Renaud Burdin to "P. Hilaire," Oct. 14, 1907, AG 2OS-96.

64. Pesse to Bailly, March 11, 1908, AG 2OS-107.

65. Pesse, Nov. 3, 1908, "Rapport sur le college français de Worcester, (Etats-Unis)," AG 2OS-115.

66. Renaud Burdin to Emmanuel Bailly, Sept. 15, 1907, AG 2OS-95.

67. Pesse to Bailly, Sept. 14, 1908, AG 2OS-110.

68. Pesse to Bailly, June 16, 1906, AG 2OS-67; Ernest Baudouy, "Carte de la Visite," typescript, April 27, 1908, 5, AG 2PK-19.

69. Pesse to Bailly, Feb. 3, 1907, AG 2OS-80.

70. Pesse to Bailly, Feb. 3, 1907, AG 2OS-80; Omer Rochain to Bailly, Nov. 25, 1908, AG 2OS-129.

71. Marie-Joseph Laity to Bailly, Aug. 15, 1907, AG 2OS-101; Pesse to Bailly, Jan. 10, 1907 [1908], AG 2OS-79.

72. Ernest Baudouy, "Carte de la Visite," April 27, 1908, AG 2PK-19.

73. Pesse to Bailly, Nov. 22, 1908, AG 2OS-111.

74. Omer Rochain to Bailly, Nov. 25, 1908, AG 2OS-129.

75. Ernest Baudouy, "Carte de la Visite," June 16, 1909, AG 2PK-22.

76. Moquin and Richards, *Assumptionists in the United States*, 57.

77. Gayraud to Odilon Dubois, Oct. 12, 1946, AC. Fr. Gayraud apparently

meant to identify Fr. Chaboud as the common superior for North America. It is also possible that he meant to write "Fr. Emmanuel, Superior general."

78. Emmanuel Bailly, "Carte de la Visite," Sept. 29, 1910, AG 2PK-23 (emphasis in original).
79. Pesse to Bailly, March 11, 1908, AG 2OS-107.
80. Gayraud to Odilon Dubois, Oct. 12, 1946, CA; Mathiot to Dubois, Feb. 19 [?], 1946, AC.
81. Mathiot to Dubois, Feb. 19 [?], 1946, AC.
82. Mathiot to Dubois, Feb. 19 [?], 1946, AC; *Prospectus*, (Worcester, [1911]), AC.
83. Baudouy, "Carte de la Visite," June 16, 1909, AG 2PK-22.
84. Bailly, "Carte de la Visite," Sept. 29, 1910, AG 2PK-23.
85. Rochain to Stéphane Chaboud, Nov. 24, 1910, AP.
86. Rochain to Stéphane Chaboud, Nov. 24, 1910, AP; "Archives de la maison," I, 12, AP.
87. Rochain to Stéphane Chaboud, Nov. 24, 1910, AP.
88. Rochain to Bailly, Nov. 29, 1910, AC.
89. "Archives de la maison," I, 12, 13 AP; Armand Goffart, A.A., "Journal d'un petit Belge," (3 parts), typescript, 2, 1, AP.
90. Rochain [?], "Dettes consolidées," n.d., AG IH-15.
91. Rochain [?], "Dettes consolidées," n.d., AG IH-15; Rochain to Bailly, Nov. 29, 1911, AC.
92. Rochain to Bailly, March 12, 1911, AC.
93. Financial report, Aug. n.d., 1911, AG 2PK-28.
94. Over six decades later, Fr. Aurèle Odil placed this discussion in 1913. However, the increase in fees to which it refers occurred in 1912. (Odil to Wilfrid J. Dufault, Feb. 10, 1976, AP.)
95. "Archives de la maison," I, 13, AP; Rochain, "Rapport semestriel," May 30, 1913, AG 2PN-70; Rochain, "Rapport semestriel," Nov. 13, 1913, AG 2PN-71; "Archives de la maison," I, 151, AP.
96. Rochain, "Rapport semestriel," May 30, 1913, AG 2PN-70.
97. Rochain, "Rapport sur la maison de Worcester (Etats Unis)," n.m, n.d. 1912, 5, AG IH-7; Rochain to Bailly, Dec. 10, 1915, AG 2PL-39.
98. Odil to Dufault, Feb. 10, 1976, AP; Rochain to Bailly, Oct. 18, 1912, AC.
99. [Rochain], "Règlement du College de l'Assomption," n.d. [1915], AG 2PN-152.

100. Rochain, "Rapport semestriel," June 8, 1910, AG 2PN-69; "Archives de la maison," I, 13, AP; Rochain, "Rapport sur la maison de Worcester (Etats Unis)," Sept. 27, 1912, AG IH-7.

101. Rochain to Bailly, Feb. 10, 1913, AC; Rochain, "Rapport sur la maison de Worcester (Etats Unis)," Sept. 27, 1912, AG-IH 7.

102. Rochain to Bailly, Oct. 18, 1912, AC.

103. Rochain to Bailly, April 7, 1914, AG 2PL-3; Rochain to Bailly, Dec. 10, 1915, AG 2PL-39.

104. Rochain, "Rapport semestriel," May 30, 1913, AG 2PN-70; Rochain to Bailly, Dec. 10, 1915, AG 2PL-39.

105. Rochain to Bailly, Oct. 29, 1914, AG 2PL-7; Rochain to Chaboud, Aug. 10, 1914, AG 2PL-8.

106. Rochain to Bailly, Oct. 29, 1914, AG 2PL-7.

107. Rochain to Bailly, March 24, 1916, AG 2PL-68.

108. Rochain to Bailly, March 31, 1916, AG 2PL-69.

109. Rochain to Bailly, June 16, 1916, AG 2PL-72.

110. Rochain to Bailly, Oct. 29, 1914, AG 2PL-7; March 24, 1916, AG 2PL-68; March 31, 1916, AG 2PL-69; June 16, 1916, AG 2PL-72; Dec. 8, 1916, AG 2PL-77; Aug. 24, 1917, AG 2PL-109.

111. Mathiot, "Histoire Ancienne," 9-10, AC.

112. Baudouy, "Carte de la Visite," June 16, 1909, AG 2PK-22.

113. Pesse, "Rapport semestriel," Jan. 10, 1908, AG 2PN-66.

114. Rochain, "Rapport semestriel," May 30, 1913, AG 2PN-70.

115. Baudouy, "Carte de la Visite," April 27, 1908, AG 2PK-19.

116. [Rochain], "Etat du College" [1918], AG 2PK-40.

117. Rochain to Bailly, Dec. 8, 1916, AG 2PL-77.

118. Rochain to Bailly, April 17, 1917, AG 2PL-105.

119. Pesse to Bailly, Oct. 28, 1905, AG 2OS-46; Rochain to Bailly, March 12, 1911; May 17, 1911, AC.

120. Claire Quintal, *Herald of Love: Father Marie-Clement Staub, A.A., Apostle of the Sacred Heart and Founder of the Sisters of Saint Joan of Arc, 1876-1936,* (Sillery, Quebec, 1984), 122-23.

121. Goffart, "Journal d'un petit Belge," 2, 4, AP.

122. "Archives de la maison," I, 7, AP.

123. *Worcester Directory 1913* (Worcester, 1913), insert following p. 32.

124. Rochain to Bailly, March 11, 1915, AG 2PL-31; Rochain to Bailly, Dec. 8,

1916, AG 2PL-77.

125. Rochain to Bailly, Oct. 1, 1915, AG 2PL-37; Rochain to Bailly, Aug. 24, 1917, AG 2PL-109; *Prospectus*: 1917-1918 (Worcester, 1917), 14.

126. Rochain to Bailly, Dec. 8, 1916, AG 2PL-77.

127. Rochain to Bailly, Dec. 8, 1916, AG 2PL-77.

128. Rochain to Bailly, Dec. 8, 1916, AG 2PL-77; Rochain to Bailly, Dec. 8, 1916, AG 2PL-77; Odil to Dufault, Feb. 10, 1976, AC.

129. Rochain to Bailly, Dec. 8, 1916, AG 2PL-77.

130. Rochain to Bailly, Dec. 8, 1916, AG 2PL-77; Odil to Dufault, Feb. 10, 1976, AC.

131. Rochain to Bailly, Dec. 8, 1916, AG 2PL-77.

132. Odil to Dufault, Feb. 10, 1976, AC; Rochain to Bailly, Dec. 8, 1916, AG 2PL-77.

133. Charter of Incorporation, AC; Goffart, "Journal d'un petit Belge," 2, 5, AP.

134. Odil to Dufault, Feb. 10, 1976, AC.

135. Rochain to Bailly, April 17, 1917, AG 2PL-105; Rochain to Stephane Chaboud, March 1, 1918, AG 2PL-141.

136. "Archives de la maison," I, 151, AP.

137. *Prospectus* 1917-1918, 7-8.

138. [Rochain], "Etat du College" [1918], AG 2PK-40.

139. "Archives de la maison," I, 151, AP; *Prospectus:* 1917-1918, 7-8; [Rochain], "Etat du Collège" [1918], AG 2PK-40; *Prospectus:* 1917-1918, 7.

140. [Rochain], "Etat du College" [1918], AG 2PK-40.

141. [Rochain], "Etat du College" [1918], AG 2PK-40.

142. "Archives de la maison," I, 20, AP; Rochain to Bailly, Aug. 24, 1917, AG 2PL-1-9.

143. "Archives de la maison," I, 20, AP; Rochain to Chaboud, Feb. 5, 1918, AG 2PL-140; "Oeuvre de la Souscription au Collège de l'Assomption de Worcester," AC.

144. "Oeuvre de la Souscription au College de l'Assomption de Worcester," AC.

145. Rochain to Joseph Maubon, Oct. 20, 1918, AG 2PL-136.

146. "Archives de la maison," I, 21, AP (emphasis in original).

147. "Archives de la maison," I, 21 AP (emphasis in original): *L'Assomption* (May 1940), 100.

148. *Prospectus:* 1917-1918, 24.

149. Rochain, "Rapport sur la maison to Worcester (Etats Unis)," Sept. 27, 1912, AG IH-7; Rochain, "Rapport semestriel," Nov. 13, 1913, AG 2PN-79; Bailly, "Félicitations et Encourgements," *Vers l'Idéal*, No. 7 (May 1913).

150. "Archives de la maison," I, 20, AP.

151. Goffart, "Journal d'un petit Belge," 2, 5, AP.

152. "Archives de la maison," I, 20, AP; Goffart,"Journal d'un petit Belge," 2, 5, AP; Régnier to Joseph Maubon, Sept. 8, 1918 [1919], AG 2PL-144.

153. Odil to Bailly, June 6, 1915, AG 2PL-47; Cordier to "un Religieux de Rome," June 21, 1905, AG 2OS-44; Rochain to E Bailly, April 7, 1914, AG 2PL-3.

154. Rochain to Bailly, Oct. 1, 1914, AG 2PL-6; Rochain to Maubon, June 21, 1918, AG 2PL-135; Rochain to Maubon, June 21, 1918, AG 2PL-135.

155. Rochain to Maubon, Oct. 20, 1918, AG 2PL-136; Regnier to Maubon, Sept. 8, 1918, AG 2PL-144.

156. Rochain to Maubon, Oct. 20, 1918, AG 2PL-136; *L'Assomption,* Series 6, No. 5 (October 1929): 65.

157. Rochain to Maubon, Nov. 29, 1918, AG 2PL-137; "Archives de la maison," I, 21, AP; Rochain to Maubon, Oct. 20, 1918, AG 2PL-136; Odil to Dufault, Feb. 10, 1976, AC.

158. Rochain to Maubon, Nov. 29, 1918, AG 2PL-137.

159. Rochain to Maubon, Jan. 31, 1919, AG 2PL-153.

160. Rochain to Maubon, April 28, 1919, AG 2PL-156.

161. Rochain to Antonin Coggia, July 17, 1919, AG 2PL-157.

162. Rochain to Bailly, Aug. 24, 1917, AG 2PL-107; Rochain to Antonin Coggia, July 17, 1919, AG 2PL-157.

163. Rochain to Antonin Coggia, July 17, 1919, AG 2PL-157.

164. Hermés Fuchs to Maubon, July 23, 1919, AG, 2PL-162.

165. "Visite de P. Joseph Maubon à Worcester (1920)," Nov. 8-13, 1920, AG 2PK-47.

166. Rochain to Antonin Coggia, July 17, 1919, AG 2PL-157.

Chapter Two
Development, Division, Depression (1919–1930)

By the end of World War I, the Assumptionists in Worcester were in virtually unanimous agreement that the college represented a magnificent opportunity for the Congregation to expand its influence and, in so doing, to extend the Kingdom of God. One of the most optimistic was the new superior, Fr. Marie-Louis Deydier. A native of France, Fr. Deydier had pronounced his religious vows in Turkey and worked there as a teacher before completing his studies in Jerusalem, where he was ordained in 1900. He taught at the Assumptionist novitiate at Pharanaki in Turkey and then was assigned to England. Transferred to Worcester from a parish in the Brockley section of London, the 47-year-old Fr. Deydier believed that "the future... for the Church as for our Congregation (outside of France) is to be found in America and in the immense British Empire." The Assumptionists, in his view, should pay less attention to the "sterile fields" of Eastern Europe and the Middle East and concentrate on the "fertile plains" of North America. Using his work in England as an illustration, he noted that in a dozen years "the single house of Brockley" had welcomed more Protestant converts into the Church than "all of our Fathers in the East received [Orthodox] schismatics in 50 years."[1]

The Need for American Assumptionists

Fr. Marie-Louis Deydier, A.A.

"We can create a unique intellectual center for the million and a half Canadian immigrants in the United States," Fr. Deydier wrote to Vicar General Joseph Maubon, but only if the Congregation would make a major commitment to staffing Assumption College with men who spoke English well and could teach natural science courses. "We have the privilege of awarding the baccalaureate, but we need professors to prepare the students for these exams," he insisted. The problem could be addressed initially by sending three or four men from France to the University of London to learn English while taking courses in the sciences, all in preparation for teaching in Worcester.[2] If, in the long run, the Congregation was to "assure the future of this beautiful work," it would have to recruit young Americans.

To do that on a significant scale, Fr. Deydier was convinced, the Assumptionists would have to establish a novitiate in America. "Be quite assured that this is the only means of assuring the future of the college," he told Fr. Maubon. "If we do not recruit here, we will not be able to continue this work." He urged that the novitiate be established immediately and be ready to receive novices by September 1920, when four or five Assumption College students would be ready and willing to enter. "You will tell me I am in great haste," he wrote. "I reply that we are at a critical moment in the work. It is time for firm and rapid decisions." He reminded the vicar general of the competition the Assumptionists were facing from other French-speaking congregations, such as the La Salette Fathers, "who vegetated in France (but) have developed themselves here in a marvelous fashion.... If we do not move quickly we will be outrun."[3]

The issues of English, science, and potential Assumptionist recruits were intertwined because, in the absence of competent

Assumptionist professors, the English and science courses were being taught by laymen, two of whom were Protestants.[4] "The lay professors," Fr. Deydier wrote to another superior, "thwart our influence most of the time even without wanting to."

The revolving staff of Europeans was also part of the problem. "We cannot maintain and develop the college with a constantly changing staff that comes from France," the new superior argued, but if an American novitiate were to be established, the Worcester community could "in a few years become self-sufficient." The five young men asking to enter the Assumptionist community, he believed, could already be supplemented by three others. "Are you going to sacrifice these vocations?"[5]

Unhappily for Fr. Deydier, the frustrations that had dogged his predecessors persisted. During his four years as superior he could not remedy the absence of an American novitiate, the dearth of Assumptionists with mastery of English or science, or the unpredictability of the flow of European personnel in and out of Greendale. Like his predecessors, he would express ardent but futile exasperation at the fact that the men reporting for duty from Europe usually arrived weeks after the school year had begun.

He was no exception. Fr. Deydier had begun his tenure as head of the college during the most disorderly inauguration of an academic year in the short history of the school. One hundred eighty students, more than the building could comfortably house, had reported for classes on September 15, 1919. In the absence of their superior, Fr. Omer Rochain, the staff of the college was presided over by the prefect of discipline, Fr. Pierre-Célestin Regnier. Two days after classes had begun, a letter from New York announced that the superior had been replaced by Marie-Louis Deydier, who was somewhere in transit from his previous assignment in England.[6]

On October 12, two returning professors who had been on "a long voyage" reported for assignment. On October 23, the new superior finally reached Greendale, along with three new professors. Although he had praise for the men who had launched the school year despite wholly inadequate staffing, Fr. Deydier found the situation "deplorable." He reorganized the classes and teaching assignments, but he could do that only provisionally because he knew six more religious would be arriving a month later. Meanwhile, other men were leaving; four days after the superior's arrival, two veterans departed to take up new assignments. On November 24, the rest of the anticipated staff, three priests and three brothers, arrived from Europe.[7]

The school year was ten weeks old before the shuffling came to an end. Six of the sixteen Assumptionists who had been present at the 1919 graduation ceremony were gone. Fourteen new men, including three returning war veterans, had taken up posts in the college, bringing the total to twenty-four religious, including seventeen priests, three choir brothers (future priests), and four lay brothers.[8]

Despite this disruption, "the future of the College looks very brilliant," the new leader wrote to his European superiors in November 1920. But success would depend "on the condition that we form professors of English and of the sciences and that we expand the buildings in order to house 300 boarding students." Before seeking a bank loan to finance further building, Fr. Deydier paid off the college's $60,000 debt to the Assumptionist Congregation in Europe. To accomplish that feat, he drew on an $8,000 loan from the Assumptionists in New York and, apparently, on some of the proceeds of the fund drive conducted by the Franco-American Catholic Federation in 1918. Once the debt to Europe had been settled, the college still owed, in addition to the $8,000 loan from New York, a $52,500 mortgage loaned by a Worcester bank to finance the 1911–12 extension.[9]

Fr. Deydier planned his new extension of the building with Onesime E. Nault, the architect who had designed the original Baltimore Avenue building and its first extension. The original structure had become a "wing" attached to the 1911–12 edifice, which had been designed by Nault to serve as the central portion of a larger whole. The objective this time was to make the central portion truly central by adding the missing northern wing. The addition would bring the capacity of the college to around 225 students rather than the 300 Fr. Deydier had at first envisioned.

While the college's real estate was being appraised for a new mortgage application, a priest from Manchester, New Hampshire, offered Fr. Deydier a gift in the form of an annuity loan of $5,000. The college would have to pay the donor 8 percent per year until his death, but would never have to repay the principal. Such gifts had come in before, always from priests. In late 1920, the college was paying interest of $650 per year on annuity loans worth $13,000. Included in that total was $2,000, at 5 percent, received from Fr. Charles Crevier, the first mover of the aborted Granby foundation, now reconciled with the Assumptionists and the college.[10]

The cost of the new wing was estimated at $90,000. Fr. Deydier anticipated having to borrow $40,000 of that from a bank. That amount, added to $24,000 the college still had on hand from the 1918 Franco-American campaign, would bring the available resources to $64,000. The superior hoped he could raise the rest by operating the college at a profit and by taking in donations from the Franco-American clergy.[11] When he set out on that quest in the spring of 1921, Fr. Deydier discovered just how deeply the French-speaking priests of New England were counting on the success of Assumption College.

After meeting with fewer than a quarter of the pastors, the superior began to think that he might have to borrow only $20,000 from the bank. A few months later, he was allowing himself to

College in 1922, with northern wing added

hope that he might dispense with a bank loan altogether. Things were going so well he might even raise more money than he needed for the new wing, so he asked permission in that case to expand the college's small farm. By August, he had brought the total up to $93,500 without recourse to a bank. Around $30,000 had come in the form of new annuity loans, the rest as outright gifts.[12]

Work on the addition began late in June of 1921, and the completed building was blessed by Bishop Thomas M. O'Leary on November 7, 1922. Besides putting up the wing and expanding the farm, Fr. Deydier had updated the whole heating and ventilation system, modernized the laboratories, rebuilt the main entrance to display "a mosaic of marble, which makes an excellent impression," installed a large three-section refrigerator, put in a hydraulic elevator, built a refectory for the nuns who staffed the kitchen, installed a large pantry and a cellar, drained swampy areas of the campus, and piped a water supply to the farm. In all, the improvements cost over $115,000, all raised by the superior without the help of a bank loan.[13]

Photo taken at 3 a.m., March 24, 1923

Fire

Fr. Deydier, his staff, and his students had only four months to enjoy the full fruits of that success. On March 24, 1923, at 2:50 on a Saturday morning, a fire alarm brought the sleeping community to its feet. The blaze was in the elevator shaft at the center of the college. With the use of the building's own firefighting equipment, the flames were prevented from spreading horizontally, but they shot up the shaft to the central roof under which the new physics and chemistry laboratories were located. At the very center of the building, the fire climbed even higher, to the roof of the tower. On a lower floor, one hundred or more of the youngest students stumbled out of the large central dormitory and made their way down two smoke-filled staircases that were already beginning to burn.

Unable to save the roof or the laboratories, firefighters protected the two wings and hosed down the central dormitory. Water cascaded through the lower floors and walls to the basement, damaging along the way the living quarters of the religious, the chapel, the library, and the study and recreation halls. By around 7:00 in the morning, the fire was out. The tower, topped by a cross, had partially crumbled, while the adjacent roof and the laboratories had collapsed into the dormitory. The students were taken in by neighbors, offered some breakfast, clothed, and sent home to their families. Easter vacation, scheduled to start the following Thursday, began dramatically on that Saturday morning.[14]

Remarkably, the students were able to return from that vacation, as scheduled, on April 9. Thanks to the labor of the Assumptionists, Assumption students who lived in the Worcester area, the college's many friends, and hired workers, it was possible, though extremely inconvenient, to reoccupy the damaged building. The school year continued and concluded without further interruption. When students returned the following September, all traces of what could have been a catastrophe had disappeared. The $100,000 in damages had also been covered, thanks to $48,000 in insurance, to the reserves Fr. Deydier had accumulated, and to another outpouring of help from the Franco-Americans of New England. "What is most astonishing," Fr. Léocard Bauer wrote back to Europe, "is the explosion of sympathy this accident has brought us and how much, deep-down, people appreciate what we are doing."[15]

The college's ordeal by fire was not quite over. In June, the renovated farm also burned. No livestock was lost, since the pigs had been sold a few days earlier and the college's four cows and two horses had been turned out to pasture. Farm buildings were lost, however, along with many chickens and pigeons. Fr. Deydier reported that only $2,000 of the $12,000 cost of rebuilding the farm would be covered by insurance. Nevertheless, he anticipated that, by the end of 1923, he would have that covered and have around $35,000 in cash, enough to start thinking about a $70,000 gymnasium.[16]

Tower and nearby rooms after the fire

Marie-Louis Deydier was meeting the college's financial challenges; he was working hard to improve the science curriculum and facilities; he saw more qualified students applying than the institution could hold; he was working on strategies for ending the college's continuing resort to "subterfuges to hide what we are lacking from the English point of view."[17] While this promising career was unfolding, however, the superior was also the central character in a period of profound discord within the community of Assumptionists stationed at the college.

To a significant extent, the dissension had local and personal origins. In time, however, it became associated with conflicts affecting the entire Assumptionist Congregation. Upon taking up the reins in Greendale, Fr. Deydier had revealed an inclination to speak and act as if nothing of great significance had been accomplished in the fifteen difficult years preceding his arrival. Some of the veteran teachers found him to be "totally lacking in delicacy, in tact, and in adroitness in his governing of the house."[18] He also had an aggravating habit of using his own experiences and the ways of the English as models for others — an approach Frenchmen, Canadians, and Franco-Americans found insulting.[19] As time passed, feelings were further embittered by Fr. Deydier's tendency to draw a sharp distinction between "good religious" and those who questioned or criticized him.

These local, interpersonal conflicts might have been contained, perhaps even resolved, had they not occurred in the context of a much larger challenge to the Assumptionist congregation and to its traditional leadership. As unlikely as it must have seemed to people familiar with the thinking and the career of Emmanuel d'Alzon, the central authorities of the Augustinians of the Assumption were embroiled in a dispute with the supreme authorities of the Church in Rome.

Discord Between Assumptionists and the Vatican

The disagreement with the Vatican stemmed from unfinished business. Fr. d'Alzon's founding of the Assumptionists had received Vatican approval in 1864, but the Congregation's constitutions had never been submitted for the formal approval required by Church law. Fr. d'Alzon himself preferred to let the constitutions evolve from experience. He did act on a constitutional issue that would later become a persistent point of discussion within the Congregation: He created more-or-less autonomous provinces. The emerging provincial system had been put aside, however, by Fr. d'Alzon's successor, Fr. François Picard, as he led the Congregation's resistance to attacks from French authorities.[20]

The governance issue did not go away. Individual members of the Congregation complained to Rome in 1909 and 1911, but Pope Pius X decided to let the ruling Assumptionists delay submission of their constitutions until a revised code of canon law had been published. The long-awaited code, promulgated in 1917 and effective in 1918, required that, for administrative purposes, religious congregations like the Assumptionists be divided into provinces.

The official Assumptionist response was predictably cool. When Superior General Emmanuel Bailly died on November 23, 1917, the code had just been published. Under the leadership of Vicar General Joseph Maubon, the ruling elders of the Congregation remained strongly committed to the centralized authority of the office of superior general. By 1918 they were under pressure from brother Assumptionists and from the Holy See to develop and submit revised constitutions. In February, a new pope, Benedict XV, ordered them to summon a General Chapter that would draft constitutions and submit them for Vatican approval. Only after that approval would another General Chapter, chosen in accordance with the new constitutions, be allowed to elect a new superior general. In the meantime, Joseph Maubon would continue as vicar general.

The Assumptionist leaders again dragged their feet. The pope himself bluntly told Fr. Ernest Baudouy — who as procurator-general handled the Congregation's relations with the Vatican — that authority among the Assumptionists was simply too concentrated. Reluctantly, Vicar General Maubon summoned the mandated General Chapter to meet in France in April 1918, while the Great War was still raging.

The General Chapter thus assembled proposed a life term for the superior general, maintenance of a central treasury, and a division of the Congregation into vague "regions" rather than well-defined provinces. In what may have been a gesture of compromise, the chapter did provide for elections in the future of delegates to regional chapters that would, in turn, elect delegates to a General Chapter. The proposed new system was submitted to Rome in December 1918, along with another plea from Procurator General Bauduoy for the maintenance of central authority over personnel and property. The Vatican, though it had prodded the Assumptionists to act, showed no haste in reviewing the Congregation's work.

Not until January 1921, did the Sacred Congregation of the Clergy finally speak. It called for further revisions in the Assumptionist constitutions, to precisely delimit the powers of the central authorities and to set up provinces under superiors with equally clearly defined powers. The Assumptionists' headstrong leadership tried once again to press its centralist case. This time, it provoked a Vatican decree, issued in June 1921, that mandated the election of a General Chapter to draft constitutions that would conform to the Vatican's directives. The chapter was to meet in Rome before the end of the year, and it would be presided over by a representative of the pope.

This dramatic intervention by the Vatican set the stage for political activity throughout the Congregation. The opportunity to elect delegates to regional chapters, delegates who would in turn send representatives to a momentous General Chapter under papal supervision, pitted the religious against one another along a number of fault lines. Reformers lined up against traditionalists, centralists against decentralizers, younger men against older ones, Frenchmen against religious who were native to, or working in, places like Belgium, England, Spain, Bulgaria, the Netherlands, Greece, Chile, and North America.

Following the elections, the second of the mandated General Chapters of the Assumptionists convened in Rome. It opened in December 1921 under the presidency of Benedict XV's personal emissary, Dom Maur Etcheverry. After some inquiry, the pope's representative concluded that there had been irregularities in the election of the twenty-five delegates. Two days later, the Sacred Congregation of Religious declared that the gathering could not under those circumstances be recognized as a legitimate General Chapter. Rather than send the assembled Assumptionists back to their homes, however, the Sacred Congregation transformed them into a "pontifical consultative commission" whose task was to bring the Assumptionist constitutions into conformity with canon law. The Sacred Congregation injected itself even more deeply into the Assumptionists' affairs, reserving to itself the right to choose the general officers of the Congregation, once the constitutions had been drafted, submitted, and approved.

During an unusually cold December and January in Rome, the "commission" carried out its task, in the process partitioning the Congregation into four provinces, three in France and one covering Belgium and the Netherlands. Three outlying vicariates would be supervised by the French provinces: South America by Bordeaux; the "East" (Eastern Europe and the Middle East) by Lyon; and North America by Paris.

Before leaving Rome, the delegates were directed to submit individually signed lists of their choices for superior general and the other seven central offices. The same directive was sent to all 365 members of the Congregation who had taken perpetual vows. After yet another year of delay, the Vatican on January 30, 1923, approved the constitutions and named as superior general Fr. Gervais Quénard, a Frenchman who had spent much of his career in Bulgaria. In the decades to come, thanks to the quality of Fr. Quénard's leadership, many an Assumptionist could readily interpret the Vatican's intervention in the governance of the Congregation as a providential blessing. Widely beloved and repeatedly reelected, Fr. Quénard would serve as superior general until 1952, when he would be replaced at his own request.[21] His successor would be Fr. Wilfrid J. Dufault, a 1929 graduate of Assumption College in Worcester.

Fr. Wilfrid Dufault, A.A. with Fr. Gervais Quénard, A.A. in 1948

A young Dufault had arrived at the college from the nearby town of Spencer during the climax of the post-war crisis among the Assumptionists. He matriculated as a second-semester high-school freshman in February 1922, a few months after the hotly contested election to the regional chapter. Among Assumptionists at the college, by that time, what had begun in 1919 as an outbreak of local discord had become entangled with the conflict that was bedeviling the Congregation as a whole.

International Discord Reaches Greendale

As noted earlier, the local disagreements at first seemed to involve only issues peculiar to the community in Greendale. In January 1920, only months after Marie-Louis Deydier's arrival, bitter complaints about the new superior's leadership were being addressed to the common superior in New York, Fr. Tranquille Pesse, and through Fr. Pesse to Vicar General Maubon. Apparently unaware of those communications, Fr. Deydier in July 1920 pronounced himself "very well pleased with all of the religious. The spirit is excellent, the work is considerable, the piety goes well, the Office is recited very regularly, the missionaries are full of zeal."[22]

However, Vicar General Maubon made a canonical visit to Greendale in November 1920, and by the time he left, Fr. Deydier had been made acutely aware of the internal criticism directed at him, which he summarily rejected as unfounded. In the next months, Fr. Deydier identified a "trinity" of men he regarded as chronic complainers. By May, one member of the dissatisfied group had been transferred to New York, but according to Fr. Deydier the remaining "recalcitrant two" were telling people they would "blow up" their superior. "All of this," Deydier declared to the vicar general, "has occasioned in me a great moral suffering and makes me detest the charge of superior. What a relief I would experience if you were inspired to place me back in the ranks."[23]

In June 1921, as the Vatican was about to issue its election decree, and as Fr. Deydier was discovering the financial generosity of the college's clerical benefactors, Fr. Tranquille Pesse as common superior in New York recommended to the vicar general that Fr. Deydier, with all his faults, be kept in office, both to support the principle of religious authority and to maintain the momentum of the fund drive and building plans. Fr. Pesse agreed with Fr. Deydier and with some others in Worcester that, for the peace of the house, it would be necessary to transfer the remaining two critics from Worcester to New York, an indication that the common superior was continuing to attribute the tensions in Worcester to localized personal animosities.[24]

Fr. Pesse was not the only one sending reports to the vicar general concerning Fr. Deydier's alleged failings. Vicar General Maubon was receiving letters written by members of the Worcester religious community and even by students. As Fr. Deydier later analyzed it, the dissatisfied religious had complained to Fr. Pesse in New York, who had encouraged them to write to the Congregation's highest authority, and some teachers had encouraged their students to do the same. When the vicar general sent Fr. Deydier a compilation of the complaints, the Worcester superior was hurt. Ominously for the future of the house, he was also angry. "All of this breaks my heart," Fr. Deydier responded to Maubon, and "makes me suffer beyond all expression, but does not discourage me. On occasion I do not disdain a fight." He insistently urged the transfer of his two principal antagonists, even though that would badly disrupt planning for the impending school year.[25] Fr. Maubon took no action, however, so the two "complainers" remained in Worcester.

The larger Assumptionist drama continued to unfold. By August 1921, instructions had been received in America for the elections that were to culminate in a General Chapter in Rome. By then, Fr. Deydier had begun to reinterpret the local conflict, framing it in terms of the disagreements buffeting the whole Assumptionist Congregation. He identified himself with the Assumptionists' central leaders and his antagonists with their antagonists. "In the difficult circumstances the Congregation is going through," he wrote to the vicar general, "I am anxious to tell you again, to you and to the assistants general, of my perfect submission to the instructions you have given for the General Chapter. I love our dear congregation with all my heart and I ask God very urgently to accord to it soon the complete approbation of the Holy Church." He said he could not understand how people like "the malcontents here" could "provoke divisions at a moment so agonizing for the future of our religious family."[26]

During the first week of September 1921, the religious at the college chose their sole delegate to the regional chapter, which was to meet September 15, also at Assumption College. Their choice was Damascène Dhers,[27] a priest always described by Fr. Deydier as a reliable supporter of the superior. The precise dynamics of the event are not clear from the surviving evidence, but a sense of the tension can be gained from a letter written by one of the critics of Fr. Deydier.

"The hostility of Fr. Marie-Louis [Deydier] against what he himself called the opposition group reached its paroxysm on the occasion of the elections for the General [i.e., regional] Chapter. . . . The result of these elections was the division of the community into two groups: 8 priests on one side and 8 priests on the other, each group with its candidate. — The division was definitive." The superior's imagination, according to this witness, became "crazed by the anger and vexation of not having been elected."[28]

Damascène Dhers was perhaps selected by the equally divided Greendale community as a compromise delegate. However, following the election, the Worcester house remained so torn that the two "camps" avoided one another whenever possible. In November 1921, a second member of the dissenting "trinity" identified early in the year was, at Fr. Deydier's insistence, moved from Worcester to New York.[29]

In Rome, the General Chapter/Pontifical Consultative Commission submitted its work to the Holy See in early January 1922, after which Assumptionist houses from Chile to the Near East were waiting to hear in what direction their Congregation would be moving and who their new leaders would be. In Worcester, Fr. Deydier, familiar with the glacial pace of Vatican deliberations, was unwilling to wait for the emergence of a new order that might impose discipline on the rebellious half of his forces. "The situation is intolerable," he told the vicar general. "Please remedy it by removing them from Worcester and by dispersing them.... You have the power and the authority to do it. The longer we wait the greater will be the harm done.... I beg you, Father, help me or else discharge me from this burden that will crush me if I am left alone to fight against all these difficulties."[30]

The outcries from Greendale finally prompted Vicar General Maubon to appoint a canonical visitor, Fr. Romuald Souarn, who was sent to Worcester in late August 1922, fully vested with the powers of the vicar general to leave men at their posts or to reassign them. Although Fr. Souarn had been told things were going badly in Worcester, he reported that "the reality exceeded everything I had been told. Two completely opposed groups who observe one another, keep an eye on one another, wage war against one another. Some never speak to one another. The slightest gesture is watched, the least word is brought back, people are determined to stop at nothing, and some are talking about recourse to Rome."[31]

To support obedience and the superior, the canonical visitor decided to remove from the house three more men Fr. Deydier had identified as serious obstacles to peace. One of them would accompany Fr. Souarn to Europe, and the other two would be placed for the time being in New York.[32] A fourth man regarded by the superior as a major fomenter of discontent was left in Worcester while his replacement and his new assignment were under consideration.[33] With the tensions in their community unresolved, the superior and his staff struggled through another eventful academic year. In November 1922, the new wing of Assumption College was formally blessed. On January 30, 1923, the Vatican approved of the Congregation's new constitutions and named Gervais Quénard as superior general. Eight weeks later, fire destroyed the central portion of the college building in Worcester. The day after the fire, while Fr. Deydier was pressing heroic effort on his men to help the school recover quickly from the disaster, another Vatican decree established the new provinces of the Congregation, including that of Paris, on which the North American vicariate would depend.

Momentous Decisions

Knowing that momentous decisions were being made in Europe, in April 1923 Fr. Deydier began to seek Fr. Quénard's permission to go to Rome for a face-to-face talk about the needs of Assumption College. On May 12 he wrote, "Now I *insist* that you *permit* me to go and explain everything to you unless you are discharging me from the college. The work is so special and so complicated that you have to be informed about things that concern the College and North America. . . . I must see you before you make definitive decisions."[34]

Momentous decisions nevertheless continued to be made to organize the Assumptionist Congregation and redeploy its leaders. On May 12, Fr. Quénard named the first superior of the Paris province, Fr. Aymard Faugère. On July 10, the new provincial superior announced that Fr. Clodoald Sérieix, superior of Bethnal Green in London, would serve as vicar provincial for North America, and that he would also serve as superior of the community in Worcester. That post would make him the new head of Assumption College. Fr. Faugère thanked the men of his new province for their prayers and letters of support, from which he took "assurance that you want to devote yourselves wholeheartedly to the coming of the kingdom of Our Lord. If that is indeed the unique goal of our religious life, we will be able to turn up our noses at our egotism and we will thus suppress the difficulties that paralyze our works."[35]

When Fr. Deydier learned that he would be transferred back to his English parish at Brockley, he thanked his new provincial, both for releasing him from the burden of organizing another school year in Greendale and for "sending me back to Brockley where I worked and where I will again be welcome."[36] However, a formal expression of thanks did not mean that Fr. Deydier was pleased with the prospect of leaving Greendale after the canonical visitor had decided to remove most of his critics. He was troubled by all that was taking place without his having had the opportunity in person to place before the European superiors his case regarding the needs of the college and his case in defense of his four-year administration. Before surrendering his post, he tried once more to make both cases in writing.

From the first, Fr. Deydier's hopes for the college had depended on the Congregation's making a major effort to supply the school with English-speaking Assumptionists. The supply had been meager, and even that had been diminished as a result of transfers during the turmoil of 1921–22. As the superior was about to leave, the situation was getting worse. "You are going to remove almost every religious who knows English adequately," he told his new provincial superior. "If the proposed changes are made, only Fr. Clodoald [Sérieix, the new superior] will be able to speak English." Already the institution's failure to provide adequate instruction in English had prompted some lay professors who were being dismissed to threaten "to denounce us to the government of the state for having obtained approval under false pretenses."[37]

Once more, Fr. Deydier pointed out that the lack of English-speaking Assumptionists was prejudicial to the most important element in his program for the college's future, the recruitment of American Assumptionists. Though there was still no Assumptionist novitiate in North America, two former students of the college had already made the journey to the novitiate at Saint Gérard in Belgium, and two more were preparing to join them. Many others were interested in the priesthood, yet instruction in English, science, and social science continued to be carried on almost entirely by laymen. "Can we hand all these young men over to the often disastrous influence of lay professors?" he asked once again. "I'll let you judge."[38]

Fr. Deydier's defense of his financial administration turned into a plea to be allowed to stay in Worcester for the good of the college. He had begun his Worcester years by paying off $60,000 the college owed to the Assumptionist Congregation. He had paid for all the building and the post-fire reconstruction without increasing the approximately $50,000 mortgage debt he had inherited. That debt remained unpaid, and he still owed $5,000 to the Assumptionists in New York, but the rest of the college's debt now was made up of around $40,000 in annuity loans, on which the principal would never have to be repaid. "But if I go," he warned, "all will be paralyzed at least for a period of two years. It will take that long for Fr. Clodoald [Sérieix] to meet the priests and the other people spread out around New England." He suggested that while Fr. Sérieix filled the new office of vicar provincial for North America and also became superior of the religious community in Worcester, Fr. Deydier might be allowed to continue as director of the college. Alternatively, Fr. Sérieix might become director of the college while Fr. Deydier continued as religious superior and had "charge of continuing relations with benefactors."[39]

Fr. Deydier's support among the Franco-American clergy of New England was the hallmark of his tenure in Greendale. In fact, early in his 1923 effort to be allowed to go to Europe to speak for himself, he had spoken in a strikingly new way about the institution's purpose and its relationship to French-speaking New Englanders. Because of the school's unique bilingual character "and because this institution has as its purpose to act against the Americanization of the French Canadian immigrants in the United States," it was not easy for Europeans to understand how it functioned.[40]

Fr. Deydier's was a significantly altered formulation of the college's mission with respect to the Franco-Americans and the French-Canadians. In 1910, Fr. Emmanuel Bailly, the third superior general of the Assumptionists, had called on his Worcester priests and brothers to join forces with French-speaking natives in a great crusade to spread Catholicism across North America. Now, in the wake of World War I, Fr. Deydier was defining the college's cultural mission in terms very familiar to Franco-Americans. Instead of marching forth under the banner of Catholicism to make Americans more like themselves, the Franco-Americans, like their counterparts in Quebec, were struggling to keep themselves and their children from becoming more like their English-speaking compatriots. From this perspective, Assumption College was expected to be a crucial resource in the mobilization for cultural defense known in Quebec and among Franco-Americans as *la survivance.*

Fr. Deydier's rewording of the college's mission reflected a personal commitment to the Franco-American way of viewing the cultural and political world. Even after he knew he would be leaving Worcester, the superior urged his new provincial superior to help the college end its reliance on Europeans and to develop "a staff adapted to the mores and to the mentality of the Canadians and of the country. The greatest reproach that has been made against us in the past is that lack of adaptation, and of understanding the Canadian mentality."[41] When it became known that Fr. Deydier would be replaced, Fr. Louis D. Grenier, pastor of Notre Dame des Canadiens Church in Worcester wrote a passionate protest to the superior general. Fr. Grenier wrote as if the mission of the college was for the Franco-Americans to define and for the Assumptionists to understand and implement. He praised Fr. Deydier for embracing not only the mission he carried from France, but also one he had learned in America.

Fr. Louis D. Grenier

"Father Deydier," he wrote, "has understood the work of this college as no one else has. Having gained our confidence, he could count on new sacrifices on the part of the priests and the laity for the growth of the college in the near future. And behold, you stop him dead in his tracks. . . . He had identified himself so well with our national work! the preservation of the faith by the preservation of the language and of our distinctive character, which so few want to take the trouble to understand."[42]

Letting Outside Dissension In

Indeed, during the spring of 1923, Fr. Deydier had apparently been prepared to play a major personal role in the perennial effort of French-speaking Catholics to defend their parishes and their schools against the assimilationist tendencies of their Irish-American bishops. On May 17 of that year, seventy-five Franco-American priests from around New England had gathered at the college for the annual celebration of the superior's feast day, enjoying a banquet and performances put on by the students.

The day had involved more than fellowship and mutual support. Fr. Deydier's report on it contained a fresh and potentially dangerous argument for his being allowed to visit Europe. "It is that the Franco-American priests wish to charge me with a mission to Rome that *you* or *Fr. Romuald* would help me accomplish. This is absolutely secret." The idea had explosive potential because it came at a time when some Franco-Americans were launching an unusually strident and divisive campaign against the claims of the mostly Irish-American episcopate. Many other Franco-Americans, while less outspoken, sympathized with the dissidents' insistence on their right to preserve and protect their cultural traditions through their parochial institutions. Among those sympathizers were many priests.[43] Some who gathered that day at Assumption were apparently prepared to go over the heads of the New England bishops by appealing to Rome, and Fr. Deydier was poised to act as their top-secret emissary, with — he hoped — the help of Superior General Quénard or of the Congregation's liaison with the Vatican, Procurator General Romuald Souarn.

Though Fr. Deydier did not reveal the content of the message with which he hoped to be entrusted, the idea of this special mission arose while dissident Franco-Americans in Rhode Island were appealing to the pope and to the apostolic delegate in Washington to rein in the bishop of Providence, William Hickey. The source of the controversy was a fund drive for a projected new high school, Mt. St. Charles Academy, in Woonsocket, a Franco-American population center. Though the school was to give prominence to the French language, it would be open to all Catholic boys and under diocesan rather than parochial control.

Militant Franco-Americans considered the scheme hostile to their cultural interests and objected to Bishop Hickey's imposition of what they regarded as a "tax" on each parish, a monetary quota that had to be met by dipping into parish savings if voluntary contributions by parishioners did not suffice. In their January 1923 appeal to the pope and their May 1923 appeal to his delegate in the United States, they defended the autonomy of the French-speaking parishes and their schools as essential for the preservation of the faith and denied the authority of a bishop to take money out of French parishes for assimilationist purposes. It was during that May that Fr. Deydier reported he had been approached about a mission to Rome.

No evidence has been found to show that the departing Assumptionist superior got any further involved in the controversy that would go down in Franco-American history as the *Sentinelle* Affair, named after the militant newspaper that first appeared in Woonsocket in April 1924. Before the dispute reached its climax in 1929, the bishop of Providence and twelve parish corporations had been sued for misappropriation of funds, the bishop's right to dispose of parish assets had been affirmed by the Supreme Court of Rhode Island, moderate Franco-American leaders and their organizations had rallied to support the bishop against the militant "Sentinellistes," a delegation of the militants had carried the signatures of 15,000 supporters to Rome, sympathetic priests had been dismissed from their parishes, and sixty-two Sentinellistes had been excommunicated.[44]

As the controversy unfolded, the Assumptionists in Worcester, far from joining in the fray, did their utmost to avoid entanglement. Late in 1925, Clodoald Sérieix, Fr. Deydier's successor, reported that "the battles between Irish and Canadians or Franco-Americans, and especially between Franco-Americans themselves continue. These are divided into two parties, of which one, with 'La Sentinelle' as its organ, accuses the other of selling out to the Irish and of betraying the Canadian cause, and the other accuses the first of being revolutionary, and of attacking ecclesiastical authority. Between the two we must swim as we can and remain quiet, or neutral, which is not always easy, especially when one and the other tries to get us on their side or at least to compromise us. We get through it by sending both of them about their business."[45]

On September 30, 1923, after four turbulent years in Worcester, Fr. Marie-Louis Deydier spoke his last formal words to the Assumption College student body: "Be studious and pious," he said in his last sermon, "and you will conduct yourselves well. Draw light and strength from Jesus in Communion and from your love and imitation of the Blessed Virgin." The next morning, in an automobile loaned by his faithful admirer, Fr. Louis Grenier of Notre Dame Parish, he left Worcester for New York, where he would embark for a journey back to his former London parish of Brockley.[46]

Perils of Borrowing and Lending

Fr. Deydier was replaced by Fr. Sérieix, who had served as pastor at Bethnal Green from 1919 to 1923. Before being assigned to England in 1908, Fr. Sérieix had served the Church for a decade in Jerusalem and in Turkey. From 1915 to 1919, he had served in medical units of the French armed forces.[47]

Fr. Clodoald Sérieix, A.A.

Upon his arrival in Worcester in September 1923, Fr. Sérieix presided over an opening of the school year that left the college packed with 225 students.[48] A year later, he would report the house overpacked with 267 students, 77 at the college level. By January of 1924, he had begun putting together plans for a new building that would house the college students separately. His intent was to bring the enrollment capacity of the college to around 350 students, with college juniors and seniors housed in semiprivate rooms rather than in large, barracks-like dormitory rooms.[49] Also in his plan was the college's first gymnasium.

Due in part to slow architectural work, it was not until 1926 that Fr. Sérieix managed to submit the plans and get final permission from Paris and from Rome to borrow $250,000 for the project. By then, however, his major superiors, both the provincial superior and the superior general, had made it clear that, due to a critical shortage of new vocations, there would not be any European Assumptionists available to help staff an expanding college in Worcester. Based on that sobering reality, Fr. Sérieix scaled down his building plans.[50] The authority to borrow, however, was not diminished, a circumstance that drew Fr. Sérieix into a complicated financial situation just as the economic slowdown leading to the Great Depression was getting under way.

Learning that the provincial superior in Paris, Fr. Aymard Faugère, was looking for means to build a church, Fr. Sérieix in the spring of 1926 offered to use some of his extra borrowing capacity to take out a $20,000 loan to support Fr. Faugère's effort. The provincial superior declined that offer, but suggested that the college instead borrow $20,000 for the benefit of a different project. The Assumptionists were attempting to put up new buildings to revive the college in Nîmes, the birthplace of their Congregation. A new chapel to be built there was projected to become the resting place of the mortal remains of the founder, Fr. Emmanuel d'Alzon. On September 15, Fr. Sérieix sent Fr. Faugère the sum requested, allowing his superior to "set the terms of the contract yourself."[51] Whatever rate of interest might be established, the priests in Nîmes were hoping to pay Worcester back using a method that had been used in the past: saying masses that had been requested and paid for in the United States.

Assumptionists in New York and Worcester were much more likely than priests in England or France to have members of the faithful pay them an honorarium, usually $1, to celebrate a mass for the special intentions of the donors — so likely, in fact, that the requests often exceeded the capacity of the priests, who said one mass a day, to satisfy the demand. In such cases, the money could be transferred to other priests located where such payments were less plentiful, such as the Assumptionists in England and France. The recipient priests would receive the honoraria and say the requisite number of masses. In the case of Nîmes, the plan was that the money would stay in the United States. The religious at Nîmes would actually say the masses, while the Americans would credit the cash collected and kept as loan repayments from Nîmes.[52]

Soon after the $20,000 loan had been sent over, there came an unexpected request from the college in Nîmes for an outright gift of $10,000 from the college in Worcester. Fr. Sérieix and his two councilors, keeping the matter secret from the rest of the house, agreed to make the gift on the condition that it be specifically authorized by the provincial superior and the superior general. That authorization came in January 1927, and in May the check was sent to Nîmes, with the hope that it would "attract the blessings of Fr. d'Alzon upon our college."[53]

In the interval a new request had arrived, from Superior General Gervais Quénard himself, for the college in Worcester to take out a loan of $50,000 to help the Congregation acquire a new general headquarters in Rome. In his response, Fr. Sérieix felt compelled to explain why this request could not be honored, at least not in full. He had put aside $50,000 of his authorized borrowing capacity for emergencies that might arise. For the loan and the gift to Nîmes, he had used $30,000 of that sum. "That leaves $20,000," he wrote, "which we will send if you want it as a loan for the general house."[54] Eventually, Rome received a $40,000 loan, half of it borrowed by the College, the other half borrowed by the Assumptionist parishes in New York.[55] At this point, with $30,000 in college funds sent to Nîmes and $20,000 to Rome, the $50,000 in borrowing capacity set aside by Fr. Sérieix for urgent needs had been exhausted.

Yet in late 1928, the provincial superior asked for another $20,000 loan for Nîmes. At that point, Fr. Sériex was pressed by his own councilors to draw the line. He wrote back to Paris, explaining that the deepening economic crisis made a positive response impossible — at least immediately. "We are at a difficult moment," Fr. Sérieix wrote, "when not only can we not get our hands on any capital, but for the first time we have had to delay the payment of some current debts. The reason: the tuition payments are not coming in; the students still owe us more than $25,000 in arrears, and it is very difficult to get those paid."[56]

Fr. Sérieix expressed his personal belief that, despite these difficulties, something more might be done for Nîmes in the not-too-distant future. Tuition payments would be coming in over the next few weeks, which would allow the overdue operating debts to be met, and a twenty-fifth anniversary fund drive was in the works. Thus "if I were the only one counting, I would willingly set aside several thousands of dollars, because I understand the urgency of saving Nîmes. But there is opposition.... So I did not press the matter, all the more so because the following statement has been expressed: Nîmes will not be able to pay interest or pay back the capital, it will therefore be the College that will have to carry the whole cost. And considering the uncertain situation in which it finds itself due to the crisis in New England, it cannot take that [cost] upon itself."[57]

Fr. Sérieix then wrote a paragraph that would haunt him for years. "The best thing would be the following: Rome owes us $20,000 (or even $21,500 because $1500 [in interest] for a year and a half has been added to the principal). If Rome could put its hands on that principal, we would accept that it [Rome] pass it on to you [the Provincial Superior] in favor of Nîmes, which would thus raise the capital it needs and [Nîmes] would assume the cost of the $20,000 additional loan." Fr. Sérieix apparently made this proposal without consulting anyone in Worcester, not even his two councilors, despite the belief among some outspoken religious that "Nîmes will not be able to pay." A year later, his successor as superior in Worcester would discover that, on top of the $10,000 gift, Fr. Sérieix had allowed the college's loan to the chronically penurious college at Nîmes to grow to $40,000 plus interest.[58]

The "College" in Worcester and the Novitiate in Quebec City

The Worcester project that had set off the sequence of loans and subloans also encountered difficulties. It was not until the end of 1925, two years after the superior had begun his planning, that construction began on the special "college" structure — to be erected behind the existing building but connecting to it — and the gymnasium.[59] Applications from prospective students continued to be abundant, and despite the late construction start, plans were developed for the use of the new facilities from the opening of the next academic year. By September 1926, work had fallen so

College building after 1926, with new rear portion attached to partly underground gymnasium (left foreground)

far behind that the opening of the school had to be delayed by several days. Even at that, when 226 high-school students arrived on September 14, the only part of the new construction that was "about finished and usable" was the gymnasium.[60] When 80 college students arrived on September 20, little had improved. Except for the 16 seniors, all of the college-level students had to sleep in the gymnasium.[61] "The new building has caused us a great deal of bad blood," Fr. Sérieix reported in December. "It was necessary to occupy a part before it was completed, and successive parts gradually as the floors were laid. We were endlessly inconvenienced by the workers, and it was reciprocal. We now occupy the three upper floors and the gymnasium, which is very beautiful, but it will be another month or two before all is finished, and until then we remain a bit in the dust."[62] The new building and the gymnasium were not finished and formally blessed until November 1927.[63]

In that same September of 1926, Fr. Sérieix was also involved in the opening of another institution — the long-desired Assumptionist novitiate near Quebec City, where Franco-Americans who felt called to become Assumptionist priests and brothers could begin their formal training as members of the religious community. Until 1926, all such candidates had been sent to the seminary at St. Gérard in Belgium. None had yet returned as an ordained priest. As vicar provincial for North America, Fr. Sérieix carried out a plan that dated back to the early 1920s when his predecessor in Worcester, Fr. Marie-Louis Deydier, had begun urging that a novitiate be erected at Bergerville, now Sillery, on the outskirts of Quebec City. The location was adjacent to the mother house of the Sisters of Joan of Arc, founded in Greendale in 1914 and still led by their Assumptionist founder, Fr. Marie-Clement Staub.[64] (In Worcester, a twelve-member contingent of the sisters continued to serve as cooks and housekeepers at the college.)

When the novitiate opened on September 4, 1926, it was under the direction of Fr. Tranquille Pesse, who had been superior in Worcester from 1905 to 1909. Fr. Staub was a member of the Bergerville house, though he devoted much of his time and attention to the affairs of the neighboring Sisters of Joan of Arc. Fr. Reginald Bonnet, also a Worcester veteran, was a member of the community. The first postulant, Leopold Braun, a 1926 graduate of Assumption College, arrived on September 8, but, as in Worcester, all was not going according to schedule. The provincial superior in Paris had not yet appointed a master of novices to guide the young postulant, and others who might follow him. In December, Fr. Sérieix was still pleading with Paris to send him a master of novices, reminding the provincial that the absence of such a director was making young Braun lose a year in his religious formation: "The fathers there are very angry with both you and me and blame us for the delay and its consequences." The master of novices, Fr. Marie-Alexis Gaudefrois, arrived from France on February 18, 1927.[65]

The problem of delay at the Assumptionist novitiate was especially disturbing because, as a source of priestly vocations in general, the college continued to be remarkably successful. In 1923, Fr. Deydier could boast of thirty-four young men who had gone from Assumption into the priesthood.[66] In 1929, the college archivist listed fifty-seven. While vocations flourished, the plan to raise a corps of new Assumptionists in America was being impeded by the very problem it was intended to solve: the lack of available Assumptionists from Europe.[67]

The shortage of religious staff, Fr. Sérieix later reported, was forcing him to appoint men to duties they should not be performing, either because they were already too busy or because they were unsuited to the work. Fr. Rodolphe Martel, for example, as prefect of discipline was assigned the task of supervising the conduct of the college-level students. Taking on that fresh responsibility did not free him from his duties as subprior of the religious community, prefect of studies, director of the orchestra and chorus, and professor of rhetoric. Three other priests were in the classroom even though in the superior's opinion they were not successful teachers and should be doing other work. "The college, which has grown in numbers, has declined in quality," the superior reported to the superior general.[68] In 1928 he reported the need to to supplement the religious staff by hiring nine lay teachers, two more than the year before.[69]

Fr. Rodolphe Martel, A.A.

The expansion of the student body came to an abrupt end with the start of classes in September 1928. Rather than welcoming 300-plus students the college saw total enrollment fall to 260 — 174 in the high school, 86 in the college. The population would continue to decline, dropping below 200 by 1933; it would not reach 300 again until 1943. From the perspective of late 1928, however, Fr. Sérieix voiced confidence in the college's ability to withstand the impact of the worsening economic crisis. Preparing to leave for a General Chapter in Europe, the superior told a meeting of the religious community on December 27 that the financial condition of the institution was "*magnifique*."[70]

A month after the superior's return in April 1929, the community learned that he would soon be going back to Europe to take up his new duties as the superior of the Paris province of the Congregation.[71] In July, it was announced that the new superior in Worcester would be Fr. Crescent Armanet, who would also succeed Fr. Sérieix as the vicar provincial for North America.[72]

Celebrating Twenty-Five Years

The spring of 1929 was, despite the tightening economic situation, a time of celebration for the college, marked by two extraordinary events: in April the visit of the French ambassador to the United States, the distinguished Catholic man of letters Paul Claudel, and in May the formal observance of the twenty-fifth anniversary of the institution. Both provided an opportunity to take stock of what had been accomplished in a quarter-century of trying to balance the sometimes conflicting demands of obeying French authorities while serving the Franco-American community in New England.

For the ambassador's information, in March, Fr. Sérieix had written a brief description of the college and an estimation of its importance. Fr. Sérieix' portrayal stressed the racial character of the enterprise in Greendale. He described the Franco-Americans as "emigrants of the French race, come especially from Canada," numbering close to two million in New England, proud of their native land and wanting to preserve "its language, religion, and traditions." They had organized themselves into parishes, and it was "for them, and at the request of their priests, of their leaders, of their societies" that the college had been founded in 1904.

As people of French racial identity, Fr. Sérieix wrote, parents who wanted to preserve their ancient language and its literary treasures could not send their sons to English-speaking American schools. Similarly, as Americans, they could not find in Canadian colleges either the English and scientific instruction their sons needed, nor the training that would familiarize them with American civic life. The answer was the bilingual college in Worcester, the unique institution that opened careers in the liberal professions to the Franco-Americans "without their having to sacrifice their language and the characteristic genius of their race." By "forming an elite for it," the college gave to "the Franco-American race in the United States" a most effective means of survival ["*survivance*"].[73]

Although the visit of the French ambassador was a major ceremonial event for the entire Worcester community, Assumption College basked in the the limelight.[74] The invitation for the visit — as part of the twenty-fifth anniversary — had been extended by the superior, Fr. Clodoald Sérieix (who joined the ambassador and his wife for the train trip from New York City); the organizational details in Worcester had been woven together under the leadership of a faculty member, Fr. Lambert Saive; and when the college superior, the ambassador, and Mme. Claudel stepped from the train at Union Station on Sunday, April 14, they were greeted by a huge and enthusiastic crowd that included the principal administrators and most of the student body of Assumption College. Before leaving Worcester two days later, the Claudels would visit and be honored by City Hall, the College of the Holy Cross, Clark University, Worcester Polytechnic Institute, North High School, and a variety of Franco-American institutions: the grammar schools of Notre Dame, St. Joseph's, and Holy Name parishes, St. Anne's orphanage, St. Francis Home, and Pasteur Hospital. Their principal attention, however, was reserved for Assumption College.

On Sunday night, the Claudels attended a concert offered in Mechanics Hall by the orchestra and choir of Assumption College, made up of students and faculty. The annual evening of music had become a popular feature of Worcester's cultural life.[75] The following noon, the Claudels were present for a banquet at the college itself. In his postprandial welcome, Fr. Sérieix went out of his way to acknowledge that the college could not have prospered for twenty-five years without the collaboration of the Franco-American clergy and the support of all of the Franco-American societies of Worcester and New England. Ambassador Claudel, in turn, congratulated the French-speaking New England clergy for the work they had been doing and pointed to Assumption College as "the most beautiful jewel" of that work. He congratulated the Franco-Americans in general for showing themselves to be good citizens of "the Great Republic" without abandoning the traditions of France and Canada, and he urged them never to content themselves with purely American qualities, reducing themselves to a "banal standardization." Instead, they should continue to develop the qualities peculiar to the French race, preserve the true French genius and its ideal, and remain faithful to its language and traditions. The ambassador, his wife, and the other guests then proceeded to the gymnasium where they were entertained by student actors in a "small drama," followed by a comedy, with musical interludes from the choir and the orchestra.

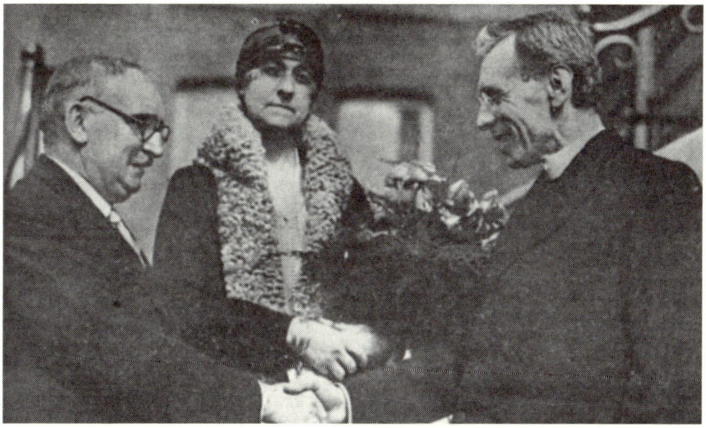

French Ambassador Paul Claudel, Mme. Claudel, Fr. Clodoald Sérieix

The next morning, the Claudels were back to say goodbye, first to the faculty and then to the students who were gathered in the largest study hall. The ambassador congratulated the young men on the theatrical and musical performances of the previous evening, urged them to continue their studies through the four college years, and spoke of the advantages of the solid education provided by a complete classical college curriculum. Invoking a privilege reserved for the most distinguished visitors to the college, the ambassador gratified the students by bestowing upon them one day of vacation.

Having loudly reaffirmed the French character of their college in April, in May the Assumptionists took full advantage of the school's silver anniversary celebration to proclaim the institution's enduring commitment to the Catholic faith in the spirit of

Emmanuel d'Alzon, whose statue on the campus was unveiled in November 1929 as part of the anniversary observances.[76] A "souvenir album" published for the occasion encapsulated the Assumptionist message as it was being conveyed to the students and to the public at the end of twenty-five years in Greendale.

Fr. d'Alzon, the album text explained, discerned in nineteenth-century France "two serious maladies." One was that "freethinkers" had multiplied and had produced many works "saying that the Christian religion with its mysteries was not reasonable, and besides, it was not possible for a god to become man." Seeing Jesus Christ as nothing more than a noble and wise human, they wished to think and live without reference to him.

The second malady followed from the first. "As faith receded, so did good morality. A God who died for us on the cross, that's something that pushes one to hate sin, to repent, and to do penance. If Jesus Christ is nothing but a superior man who was killed because he was misunderstood, that does not create any obligation at all; his life is a beautiful example: praise to those who follow him! but those who do not imitate him are not acting badly; one has the right to make of life a party filled with pleasure; one can think, as before him [Jesus], that each is the master of his own body and soul, and that goodness and happiness are whatever is pleasing to each person at whatever moment it happens to please. Thus, as faith in the divinity of Our Lord Jesus Christ disappears, every element will soon disappear of the greatness, of the virtue, of the profound joy, of the liberty, of the justice, and of the tenderness that have been deposited in the humblest of souls over the course of nineteen centuries."

Fr. d'Alzon therefore insisted that the first step that had to be taken was to restore faith in Jesus Christ. "Being God he created everything and is master of all. His commandments oblige men: every act that disobeys them is evil; any law that ignores them, or that permits what they forbid, is evil. There is but one truth, the one he preached; one virtue, the one he preached; one cult, the one he established; one religious society, the Catholic Church, which alone conserves everything he said, lives everything he did, offers it to God as he offered himself. When all men are in submission to to this Church, when all societies are governed according to its laws, then 'the kingdom of God' will have reached all the way to us. It is necessary to work for the coming of that kingdom. This is how Fr. d'Alzon saw his work."[77]

The work and the spirit of Fr. d'Alzon, the souvenir album declared, were well established at Assumption College. "In the classroom, care is taken to see that the history of the world appears above all as the history of the Church, and Jesus Christ as 'the continuance of history,' (as Fr. d'Alzon put it), and everywhere, under the wisest or the most criminal of political systems, [there appears] the mighty hand of the One who is the Father and Master of men, building the City of God, with them or in spite of them, establishing or avenging the rights of his Son, who has saved men and will return to judge them." Courses in literature at the college included study of the Gospels and the Greek and Latin Fathers of the Church. The point was not "to show that they have an art superior to, or even equal to, that of the pagan authors. But...ideas penetrate with language, and it gives students a false view of the world to make them think endlessly in terms contrary to the faith."[78]

The description of religious instruction at the college showed that the spiritual sons of Fr. d'Alzon remained committed to struggle against the forces of irreligion in modern culture. "Today it is clear, to anyone who wants to see, that it is no longer the statutes of saints, certain interior details, that are under attack, it is the very foundations of the whole religious edifice that people are trying to shake. To do that they employ all the resources of modern progress, they poison the most beautiful discoveries of the natural sciences, distorting their true conclusions to turn them against the existence of God and of the supernatural order. It is modern warfare, with its long-distance cannons, its machine guns, its airplanes, its submarines. One cannot hope to knock down the new enemies of religion with a rusty lance or with a bow and some arrows. One must adapt oneself to the needs of the times.

"That is why we mean to give to our students, by a gradation appropriate to their intelligence, a solid religious armature, capable of resisting all assaults. Especially in the higher courses, it is important to bring out certain aspects of Catholic faith and morality, certain objections, raised a thousand times and refuted a thousand times, or certain historical errors, certain rash conclusions of modern science, and to offer students a sharp, peremptory reply to the attacks of incredulity. Our professors of history, of apologetics, and of philosophy know wonderfully well all the new tactics of the enemy, and how to respond to them. The young man who leaves us at the age of 22 or 23, proudly carrying his diploma, also carries weapons to defend his religious faith."[79]

Death of the First Son

The celebratory mood of April and May 1929, tempered by uncertainty over the spreading economic distress, was shattered in July by the sudden and unexpected death of the first American Assumptionist, Fr. Louis Robert, at the age of thirty-one. Fr. Robert had entered the college from Westfield, Massachusetts, as a high-school freshman in 1915. He was one of five gifted members of that extraordinary freshman class who made such rapid progress that they had completed their freshman studies by Christmas and ended the academic year at the top of the sophomore class. In 1918, when the institution faced a severe shortage of teachers, as a college freshman he taught geometry to high-school sophomores. The following year he taught accounting and trigonometry to seniors. He also at various times taught typewriting and served as a monitor supervising student conduct. At the end of his junior year in 1921, he asked to be admitted to the Assumptionist community and became the first student from the college to make the journey to the novitiate at St. Gérard in Belgium. He completed his studies at Louvain and was ordained there on July 28, 1928. Fr. Robert was then assigned to Assumption College, where he served as the principal monitor in the high school and taught medieval history to high-school seniors. On July 3, 1929, less than a year after his ordination, he drowned in Baker Lake in Charlton, where the Assumptionists had established a summer retreat a few years earlier.

Only weeks before his death Fr. Louis Robert had preached at the alumni reunion in Greendale. He had closed his sermon with a prayer:

Louis Robert
as a student

> *"Lord, bless this house, bless the superior, the fathers and professors, the students. Bless those who have died. . . . Bless the parents of the students and the benefactors of the college.*
>
> *"This college has been good for me. May it long continue to make you known and loved by my brothers; may it always make us appreciate and love our race and our United States! I ask for only one thing: to be able in heaven, in union with all its former professors and students, in union with our mother, Our Lady of the Assumption, to praise you and glorify you forever!"*

Fr. Odilon Dubois, who maintained the handwritten journal known as the "archives of the house," after completing his account of the life, death, wake, funeral, and burial of Fr. Robert, also closed with a prayer: "Through his intercession may he continue from heaven the work he had only begun here below!"[80]

On June 17, a little more than two weeks before the drowning of Fr. Robert, Fr. Clodoald Sérieix had left the college to take up his duties as provincial superior in Paris.[81] Fr. Crescent Armanet arrived to begin his work as religious superior and president of the college on August 8, 1929.[82] He had begun his Assumptionist life in 1891 as a student in a French alumnate. Following his novitiate, he took his final religious vows in 1898. He went to Notre-Dame de France in Jerusalem to study philosophy and theology, then to Turkey for another year of study. He was ordained in Constantinople at the end of 1903 and taught in Assumptionist schools in

Turkey and Bulgaria before being assigned to parish work in England from 1902 to 1912. His mastery of English served him well when he was assigned to one of the two Assumptionist parishes in New York City from 1912 to 1914. During World War I, he was summoned home to military service, first in the paramedical corps and later with the High Command in Paris. He returned to parish work in New York until he was assigned to replace Fr. Sérieix both as superior at Assumption College and as vicar provincial for North America.[83]

Contemplating the Vitality and the Debt

The day after his arrival on August 8, Fr. Armanet took to the road to begin assessing the college's standing in the eyes of its most important supporters. By the end of August, he had called upon the mayor of Worcester; the bishop of Springfield; the *Worcester Telegram; L'Opinion Publique;* the French pastors in Worcester, Southbridge, Webster, Millbury, New Bedford, Fall River, Springfield, Chicopee, Chicopee Falls, Holyoke, Fairview, Leominster, Fitchburg, Lowell, all in Massachusetts; Woonsocket in Rhode Island; Willimantic in Connecticut; and Nashua and Manchester in New Hampshire. On August 28, he met with more than one hundred alumni gathered at the college for the feast of St. Augustine. In October, he addressed a meeting of l'Union St. Jean-Baptiste in Burlington, Vermont. He came away with accolades about the college, spoken by French ambassador Paul Claudel, ringing in his ears; with the society's pledge of $1,000 per year for the next ten years; and with a deep personal impression that "our establishment now occupies an immense place in the heart of all the Franco-Americans of New England."[84]

Fr. Crescent Armanet, A.A.

A week after Fr. Armanet wrote those optimistic words, the New York Stock Market collapsed. During his months of testing the level of public support for the college, the new superior had also been discovering the state of the college's finances. On that front, he found the prospects less heartening. Student enrollment had continued to decline and was at 240 in the fall of 1929. With so few boarding students, Fr. Armanet and his staff were thinking seriously of admitting commuting students the following year, boys who would study at the college but return to their homes at the end of the day.[85]

As difficult as the enrollment problem was, the college's accumulated debt hovered even more ominously over the new superior's view of the future. He was especially disturbed to learn of the loan that had been silently transferred from the General House to Nîmes. To Superior General Gervais Quénard he wrote: "I was very astonished to learn the other day from Fr. Clodoald that the $20,000 loaned to the General House had been passed on to the house in Nîmes, with his approval, but without the knowledge of any of the fathers here in the College!! Now, Nîmes cannot reimburse either the interest or the capital. So here we are again in the sack. Our debt is even greater than I thought. And that's saying something."[86]

He could not resist taking his anxieties out on the man who had handled the college's finances since 1923, the man who was now his provincial superior, Fr. Clodoald Sérieix. "It's nice to lend money when one is covered with debts oneself," he wrote. "Admit that you have been a bit too generous with the money of the College, which was itself over its head in debts." By Fr. Armanet's calculation, the debt, which was at $52,000 at the end of Fr. Deydier's tenure in 1923, had risen to around $350,000 by 1929.[87]

In his own defense, the provincial superior offered his calculation of the debt and its sources: Fr. Sérieix had borrowed $295,000, of which $255,000 was for the college's use in the construction; $20,000 was for the first loan to Nîmes; and the other $20,000 was for the loan to the General House that ended up in

Nîmes. In addition, when he left, the college still owed $18,000 to the contractor and $4,000 to the architect of the new buildings, for a total of $317,500. Against this he credited $18,000 in uncollected tuition and fees owed by the students, bringing the total down to "$300,000 at most."[88]

Back in Worcester, the new superior lamented, "I sometimes have bad dreams when I think about the financial difficulties of this house caused by the debt."[89]

Notes

ABBREVIATIONS
AC — Archives of Assumption College, Worcester, MA
AG — Archives of the Assumptionist General House, Rome
AP — Archives of the Assumptionist Provincial House, Brighton, MA
PA — Archives of the Assumptionist Province of Paris
SP — Archives of the Diocese of Springfield, MA

1. Robert Fortin, A.A., ed. and trans., *Windows on Assumptionist History: Short Biographies of Assumptionist Religious, 1850–2000* (n.p., 2002), 103–4. Deydier to Joseph Maubon, May 2,1920, AG 2PL-164.
2. Deydier to Maubon, May 2, 1920, AG 2PL-164; Deydier to Maubon, Jan. 27, 1920, AG 2PL-163.
3. Deydier to Maubon, Jan. 27, 1920, AG 2PL-163; Deydier to Maubon, May 2, 1920, AG 2PL-164.
4. Deydier to Maubon, Jan. 27, 1920. AG 2PL-163.
5. Deydier to Maubon, Jan. 27, 1920. AG 2PL-163; Deydier to Stéphane Chaboud, May 29, 1920, AG 2PL-167.
6. "Archives de la maison," I, 25 AP.
7. Deydier to Maubon, Jan. 27, 1920, AG 2PL-163; "Archives de la maison," I, 25–26, AP.
8. "Collège de Worcester (Etats Unis)," MS list, AC; "Archives de la maison," I, 24–26, AP.
9. Deydier, "Maison de Worcester," Dec. n.d., 1920, AG 2PK-53; Deydier to Maubon, Jan. 27, 1920, AG 2PL-166. The campaign had brought in around $65,000 (Hermés Fuchs to Maubon, July 23, 1919, AG 2PL-162). Much of the money raised was used to buy land to protect the campus from the encroaching development of Greendale. (*Souvenir de ses Noces d'Argent* [Worcester, 1929], 16). In March 1921, Fr. Deydier reported that he still had $24,000 in the bank from the "Drive" (Deydier to Maubon, March 16, 1921, AG 2PL-172); Anon., "Finances," Nov. n.d., 1920, AG 2PK-50 and 2PK-51.
10. Deydier to Maubon, Feb. 6, 1921, AG 2PL-169; Anon., "Finances," Nov. n.d., 1920, AG 2PK-50 and 2PK-51.
11. Deydier to Maubon, March 16, 1921, AG 2PL-172.
12. Deydier to Maubon, April 14, 1921, AG 2PL-173; Deydier to Maubon, July 22, 1921, AG 2PL-178; Deydier to Quénard, June 27, 1923, AG 2PL-231.

13. "Archives de la maison," I, 28, 32, AP; Deydier to Maubon, Oct. 2, 1922, AG 2PL-202.

14. Anon., *Album-Souvenir offert par le Collège de l'Assomption a tous ses amis a l'occasion de ses Noces d'Argent* (Worcester, 1929), 18; Polyeucte Guissard, A.A., "Un demi-siècle d'histoire," *Assomption/Assumption* (1954), 151.

15. Guissard, "Un demi-siècle d'histoire," 151; Anon., "Le Collège de l'Assomption," *Opinion Publique* (Worcester), April 28, 1923; Deydier to Gervais Quénard, May 17, 1923, AG 2PL-230; Léocade Bauer to Quénard, April 7, 1923, AG 2PL-234.

16. Deydier to Quénard, June 27, 1923, AG 2PL-231.

17. Deydier to Aymard Faugère, July 16, 1923, AP.

18. Tranquille Pesse to Maubon, Jan. 30, 1920, AG 2PK-46.

19. "Visite du P. Joseph Maubon à Worcester" (1920), AG 2PK-47.

20. Unless otherwise noted, the following discussion is based on Joseph Girard-Reydet, A.A., *Le Père Gervais Quénard, Supérieur général des Assomptionistes* (1875–1961) (Paris, 1967), 135–55.

21. Richard Richards, A.A., *The Assumptionists,* (New York, 1980), 60.

22. Pesse to Maubon, Jan. 30, 1920, AG 2PK-46; Deydier to Maubon, July 30, 1920, AG 2PL-166.

23. Deydier to Antonin Coggia, Feb. 28, 1921, 2PL-181; Deydier to Maubon, May 24, 1921, AG 2PL-175.

24. Pesse to Maubon, June 7, 1921, AG 2PK-57.

25. Deydier to Maubon, June 10, 1921, AG 2PL-176.

26. Deydier to Maubon, Aug. 29, 1921, AG 2PL-176.

27. "Archives de la maison," I, 29, AP.

28. Paulien Vassel to Maubon, Aug. 10, 1922, AG 2PL-224. The body of priests living in the Worcester house could not have voted in the election for the General Chapter, but they did elect a delegate to the regional chapter.

29. "Archives de la maison," I, 29, AP.

30. Deydier to Maubon, June 30, 1922, AG 2PL-199.

31. Souarn to Maubon, Aug. 28, 1922, AG NI-404.

32. Souarn to Maubon, Aug. 28, 1922, AG NI-404; Souarn to Maubon, [Aug. 30], 1922, AG NI-405; "Archives de la maison," I, 31, AP.

33. Souarn to Maubon, Aug. 28, 1922, AG NI-404; Aug. 30, 1922, AG NI-405.

34. Deydier to Quénard, April 15, 1923, AG 2PL-227; Deydier to Quenard, May 12, 1923, AG 2PL-229.

35. Aymard Faugère, circular letter, July 10, 1923, AG 2PK-71.

36. Deydier to Quénard, June 27, 1923, AG 2PL-231.

37. Deydier to Faugère, July 16, 1923, AP.

38. Deydier to Faugère, July 16, 1923, AP.

39. Deydier to Faugère, July 16, 1923, AP.

40. Deydier to Quénard, May 17, 1923, AG 2PL-230.

41. Deydier to Faugère, Aug. 13, 1923, AP.

42. Grenier to Quénard, Aug. 20, 1923, AP.

43. Deydier to Quénard, May 17, 1923, AG 2PL-230; Armand Chartier, *Histoire des Franco-Américains de la Nouvelle-Angleterre:1775–1990,* (Sillery, P.Q., 1991), 147, 149, 152, 155.

44. Chartier, *Histoire des Franco-Américains,* 143–69. All of the excommunicated discontinued their publication and formally submitted to Church authority.

45. Sérieix to Thomas Darbois, Dec. 1, 1925, AP.

46. "Archives de la maison," I, 40, AP.

47. *Lettre à la famille,* New Series, 67 (Feb. 15, 1949), 15.

48. Sérieix to Pierre-Fourrier Merklen, Sept. 13, 1923, AG 2PL- 244.

49. Sérieix to Faugère, Sept. 17, 1925, AAP; Sérieix to Faugère, May 2, 1924, AG 2PK-74.

50. Sérieix to Quénard, Feb. 3, 1926, AP.

51. Faugère to Sérieix, June 19, 1926, AP; Sérieix to Faugère, Sept. 15, 1926, AP.

52. Sérieix to Faugère, Dec. 11, 1926, AP.

53. Sérieix to Faugère, Dec. 11, 1926, AP: Sérieix to Quénard, May 19, 1927, AP.

54. Sérieix to Quénard, April 16, 1927, AP.

55. Sérieix to Quénard, April 16, 1927, AP; Sérieix to Faugère, July 3, 1927, AP; Sérieix to Quénard, Jan. 26, 1928, AG 2PK-83.

56. Sérieix to Faugère, Dec. 31, 1928, AP.

57. Sérieix to Faugère, Dec. 31, 1928, AP.

58. Sérieix to Faugère, Dec. 31, 1928, AP; Armanet to Quénard, Dec. 10, 1929, AG 2PL-283.

59. Sérieix to Rochain, Dec. 1, 1925, AP.

60. Sérieix to Faugère, Sept. 15, 1926, AP.
61. "Archives de la maison," I, 60–61, AP.
62. Sérieix to Faugère, Dec. 11, 1926, AP.
63. *Souvenir de ses Noces d'Argent,* 75.
64. Deydier to Joseph Maubon, July 30, 1920, AG 2PL-166.
65. Sérieix to Faugère, Sept. 15, 1926, AP; Sérieix to Faugère, Dec. 11, 1926, AP; *L'Assomption* 4, 10 (Oct. 1936): 234.
66. Deydier to Faugère, Aug. 13, 1923. AP.
67. "Archives de la maison," I, 67–69, AP.
68. Sérieix to Quénard, April 16, 1927, AP.
69. Sérieix to Quénard, April 16, 1927, AP; Sérieix, "Rapport semestriel," Feb. 6, 1928. AG 2PN-73.
70. "Archives de la maison," I, 78, AP.
71. "Archives de la maison," I, 89, AP.
72. "Archives de la maison," I, 8, 95, AP.
73. "Un College Francais aux Etats Unis," March 1928," AP.
74. The details of the Claudel visit are based on "Archives de la maison," I, 82–87, AP, and Francis J. Murphy, "Paul Claudel: Ambassador of French Culture," *New England Quarterly* 60 (1987): 250–55.
75. Sérieix to Quénard, March 10, 1924, AG 2PL-246.
76. *Worcester Telegram,* Nov. 25, 1929, 1.
77. *Souvenir de ses Noces d'Argent,* 21–22.
78. *Souvenir de ses Noces d'Argent,* 24.
79. *Souvenir de ses Noces d'Argent,* 36.
80. "Archives de la maison," I, 90–94, AP; *L'Assomption,* Series 6, No. 5 (October 1929): 65–67. According to a later article in *L'Assomption,* New Series 3, 1 [Jan. 1934]), Louis Robert as a college freshmen taught science courses to high-school juniors and seniors.
81. "Archives de la maison," I, 90, AP.
82. "Archives de la maison," I, 90, 96, AP.
83. "Archives de la maison," I, 90, 96, AP. Fortin, *Windows on Assumptionist History,* 14–15.
84. "Archives de la maison," I, 96, AP; Armanet to Paul-de-la-Croix Journet, Oct. 19, 1929, AP.

85. Armanet to Sérieix, Oct. 29, 1929, AP.

86. Armanet to Quénard, Dec. 10, 1929, AG 2PL-283.

87. Armanet to Sérieix, Nov. 19, 1929, AP.

88. Sérieix to Armanet, Dec. 20, 1929, AP.

89. Armanet to Sérieix, Nov. 19, 1929, AP.

Chapter Three
Depression, War, and the "Perilous Honor" (1930–1946)

Over the course of the next decade and a half, the Assumptionists struggled to carry their college intact through a severe economic depression and a military crisis that virtually emptied the college-level classrooms and dormitories. An important new element entered the college's story in the persons of young Franco-American Assumptionists who were being ordained and assigned to teach in Greendale. In time, members of this new cohort would begin to think that their turn to lead the institution was being unreasonably delayed by their superiors in Europe. As the Franco-Americans thought about the college they would eventually inherit, they became increasingly aware of yet another historical change. The Franco-American population of New England, the bedrock of the college's support, was becoming assimilated into the English-speaking culture that surrounded it. Before very long, the institution known as "the Franco-American college" might become obsolete.

In the spring of 1930, spurred by the developing economic crisis, Fr. Crescent Armanet inaugurated a practice that would become a staple of life for the Assumptionists over the next decade: personal recruitment of prospective students in the Franco-American communities. On weekends, the priests tried to find suitable students in the parishes they regularly served. During the Easter vacation — when they believed parents were making decisions about where to send their sons to school — they also visited as many of the neighboring parishes as they could reach. Fr. Armanet also went into the study halls of the college to urge students to help find new members for the student body; he promised a day of vacation to those who succeeded.[1] Beginning in 1930, the college also accepted "externs," commuting students from Worcester and neighboring towns.[2]

The fact that the Assumptionists worked without pay continued to allow the annual cost of an Assumption education to remain modest: resident students were charged $230 for room and board, $100 for tuition, $12 for bed linens, and $8 for "library and sports." Commuting students (externs) paid $100 for tuition and the $8 fee.[3] The number of students arriving for classes rose from 240 in September 1929 to 273 in September 1930, but personal effort could not for long keep up with the effects of economic decline. In fall 1931, 263 students presented themselves. There were 215 in 1932, 197 in 1933, and 197 again in 1934. By then, Fr. Antoine Philippe had been assigned to devote four days a week to the task of recruiting students.[4]

The number of externs was disappointing: nineteen in 1931–32, fifteen in 1933–34, twenty-four in 1934–35. Fr. Armanet complained in 1934 in the college magazine, *L'Assomption,* that "the Franco-American population of Worcester which amounts to around 25,000 souls has not yet understood the immense advantages we are placing within their reach in creating the externate. . . . The infatuation with the public high school that we see among so many of our people, to the detriment of our bilingual college, augers no good for the future of our race in this large city of 200,000 inhabitants. It is important to react immediately against these shameful surrenders, or else our influence, as an ethnic group, will soon be null."[5]

Also in the spring of 1934, in an attempt to boost enrollment and without mentioning that he was reviving an old practice, Fr. Armanet announced plans for a prefreshman preparatory class, intended for Franco-American boys who did not know enough French to begin regular studies. "If, instead of coming to us," he wrote, "these little Franco-Americans go to the public high school they are lost to our people and, perhaps, to our faith, because their ignorance of French will inevitably keep them away from their racial brothers and will push them little by little into the arms of the Protestants." In October, he had to report that the preparatory course was not being offered, "for lack of candidates. It's very regrettable, but there is nothing we can do about it."[6]

The Curriculum

The school, to which students were being so painstakingly drawn, continued to offer both a classical education intended to promote clarity of thought and expression, and a Catholic education designed to strengthen and deepen the faith boys carried with them from their families and their parochial schools. In his 1929 sermon to the alumni, Fr. Louis Robert recalled how frequently Assumption students in his day had questioned the utility of one or another aspect of the academic program, only to be told by professors, "*Non scholae sed vitae discimus,* (It is not for school but for life that we learn)." "Greek and Latin have perhaps long been forgotten," the young priest observed, "but from the study of those two languages we still possess the habit of thinking and of expressing ourselves with clarity; few among us probably would know today how to solve algebra problems or prove geometric theorems, but that is of little importance: we have what we sought to acquire by the study of mathematics, which is precision and exactitude. *Non scholae sed vitae discimus:* our studies were a seed from which we now harvest the fruits.

Bro. Louis Robert in 1923

"When we left the college," he continued, "we took with us for our enjoyment the qualities of spirit that we had acquired here, and now we are enjoying them. We have, for example, a greater facility for understanding people and things, which is a great thing and important in any profession. We experience a greater participation in the joys and sufferings of others. Life for us has more joys, we are less often caught unawares, and we have a greater taste for all that is beautiful."

But the day would come when, "after years of labor, the lawyers will close their books, the doctors their bags of instruments; the architects and engineers will put aside their rules and their compasses, the priests will stop giving blessings — and their eternity will begin.

"At that point the Latin they will have learned will serve them no more; they will have no use for history and sciences; geography will be useless. — And if those were all that the college had given us, it would have failed in its mission and been unjust to us.... If we miss heaven, we have missed everything!

"At Assumption we have learned not only to live, but also to die — to prepare ourselves for death, and of the benefits we have received, that is the greatest and the most beautiful; that is the only benefit that counts; without that one, all the others are worth absolutely nothing."[7]

The young priest's words echoed the message that had long appeared in the college's catalog: "The intellectual development of a student, without religion to purify his heart and guide his will, cannot produce serious results. Thus it is the constant preoccupation of the Assumptionist Fathers to promote the natural flowering of each of the student's faculties, especially those that give the student a sense of direction and make him a man of character and of duty, loyal to his faith and to his country."[8]

Student behavior was closely monitored, and religious instruction was part of every student's program of study every semester. In addition, students gathered each morning and evening in the chapel for prayer, gathered again for daily mass, heard weekly spiritual discussions, and were expected "to regularly complete [their] religious obligations, to approach the sacraments often, and to make an annual three-day retreat. Under the rules of general discipline, all mail to and from students was under administration "control." Students were forbidden to receive newspapers or other publications without special permission; books brought in had to be examined and approved; and no visitors were allowed other than parents or their authorized representatives.[9]

The enduring classical curriculum continued to place heavy emphasis on the study of languages and literature, with French as the language of instruction for many courses. High school freshmen studied catechism, Latin, French, history, and geography, all in French, and mathematics and English in English. Sophomores studied the same subjects — with American history in English — and began the study of Greek (in French). Juniors continued catechism, Latin, Greek, French, history in French, English and mathematics in English and added "general science" in English. Seniors continued all of those subjects and added civics in English.[10]

To be admitted to the college-level program, a student had to have passed at least four years of Latin, English, and French, a year and one-half of both Greek and algebra, and one year each of plane geometry, religion, and American history and civics. Those standards tended to limit admission to young men who were graduates of the high-school portion of the college.[11]

College freshmen studied religion in French, as they did Latin, Greek, French, history, and geography. Their English and physics courses were in English. The sophomore program substituted general chemistry for physics and added astronomy, also taught in English. Juniors began the intense study of Thomistic philosophy with logic, cosmology, and psychology in Latin and apologetics in French. In English, they studied American government and English, and they also took a course in Latin conversation, focusing on issues raised in the philosophy courses. In addition, juniors had to make a choice between a premedical program (general biology in English), an education program (philosophy of education and psychology of education in French) and a prelaw program (jurisprudence in French, biology and geology in English).[12]

Seniors continued philosophy (ontology, criteriology, theodicy, general ethics, special ethics) in Latin and took apologetics, history of the Church, social and political economy, and history of philosophy in French. The premedical option for seniors was general chemistry in English; the "pedagogical" option included courses in history of education, principles of secondary education, and methods, observation and practice of teaching, all in French; prelaw students studied history of economic thought and medieval institutions, both in French.[13]

College seniors who did not choose general chemistry thus took all of their courses in Latin or in French. That remained true when, beginning in 1934–35, the philosophy courses began to be taught in French rather than in Latin.[14]

All students were required to spend about five hours each day in study halls or in their rooms preparing for classes or completing written homework assignments. College freshmen and sophomores also had to write a weekly "dissertation or composition" to improve skills of expression "and to train them to think for themselves." Juniors and seniors, in addition to taking semester exams, were required to write monthly compositions on the "principal materials" of their philosophy courses and to take monthly exams on "all materials." Approximately once a month, philosophy students took turns presenting a public defense of a philosophical thesis. One student would present the thesis and two would attack it in formal arguments before the junior and senior classes as well as invited members of the faculty.[15]

To receive the bachelor of arts degree, students had to pass a special examination at the end of the senior year. The written portion was a thirty-to-fifty-page dissertation on a topic in philosophy selected by the faculty jury. Before the two-part oral exam, each student selected, subject to the approval of the jury, forty-five theses to be defended in a public session. Of these, thirty were to be

Bachelor of Arts examining board

in philosophy, five in apologetics, and five in economics or education. Before that "ordeal," the jury in a private session questioned the student on "other materials." In the public session the student had to defend his selected theses, and one of the philosophical theses had to be defended in Latin.[16] Students who failed the dissertation or either part of the oral examination were sometimes allowed to try a second time, but it was not unusual for one or two students per year to "finish" the college program without receiving the bachelor's degree.[17]

In 1932, the high-school program was approved by the New England College Entrance Certificate Board, which meant that students with an Assumption diploma would not have to take entrance examinations for admission to any of the twelve institutions that made up the association. However, despite the college-level program's rigorous curriculum and high academic standards, it was not accredited by any established association of colleges and universities. A major hindrance had long been the virtual absence of university degrees among the European Assumptionists on the faculty, which stemmed from the Congregation's history of opposition to state-controlled education in France.[18] It was widely hoped that the rising generation of college-educated Franco-American Assumptionists would contribute in a major way to the solution of the problem.

American Assumptionists in 1937. Left to right: Frs. Clair Boisvert, Henri Moquin, Leo Martin, Wilfrid Dufault, Antonio Laberge, Pierre-Rodolphe Martel (a Canadian), Ulric Charpentier, Edouard Moreau, and Armand Desautels Absent were Frs. Marcellin Parent and Leopold Braun

The Rising Generation: American Assumptionists

The time indeed had come to begin thinking about a generational change in the college and religious community. In a report written early in 1928, Fr. Clodoald Sérieix had reminded his superiors across the Atlantic that the European Assumptionists, though they had been embraced by the Franco-Americans, were nevertheless regarded in New England as outsiders. "Even though we've done them an invaluable service and they appreciate it, even though they esteem us as the only French religious who have understood their mentality and have remained faithful to them, nevertheless *to them we are foreigners*; and even our best friends, with few exceptions, long for the day when in the place of Frenchmen, it will be

Canadians or Franco-Americans who have the College. We are here only to prepare the place for these last."[19]

In the early 1930s, a small but very important stream of Franco-American Assumptionists began to return to the college as ordained priests. They were the harbingers of the changes Fr. Sérieix had anticipated just two years earlier. All parties agreed that these men, who held Assumption College bachelor's degrees, should be encouraged to pursue master's and doctoral degrees. The 1933–34 faculty included three Franco-American Assumptionists: Fr. Leopold Braun (professor of English, general science, and German, director of the college's symphony orchestra), Fr. Ulric Charpentier (professor of geometry, monitor), and Fr. Henri Moquin (professor of inorganic chemistry).

Another Franco-American, Fr. Edward Moreau, had already been sent to live in one of the Assumptionist parishes in New York while pursuing a master's degree in English at Columbia University. Fr. Moquin was studying chemistry at Clark University while teaching at Assumption, often putting in eighteen-hour days. As anticipated by Fr. Sérieix, they and their younger colleagues, most of whom pursued advanced studies, gradually took on a group identity as the Franco-American progeny that was to succeed the founding European Assumptionists. When the college magazine in 1937 published a photograph of "The Young Franco-American Generation of Assumption College," there were nine priests in the picture, with two others listed as unavailable for the occasion.[20]

One of those absent was Fr. Leopold Braun, for whom a different apostolate had begun. His superior, Fr. Crescent Armanet, was sent to Washington in January 1934, while the United States was in the process of establishing diplomatic relations with the Soviet Union. At the direction of Superior General Gervais Quénard, Fr. Armanet went to the nation's capital armed with two letters from James Roosevelt, son of President Franklin D. Roosevelt. His objective was to secure an arrangement whereby an American Assumptionist might enter the Soviet Union as chaplain to American Catholic personnel who would be stationed in Moscow. To his considerable surprise, Fr. Armanet could write to his provincial superior on February 11 that "our efforts, placed under the protection of St. Therese of the Infant Jesus, have been crowned with success." Within weeks Fr. Braun had left the faculty and the orchestra of Assumption College and flown to Moscow, where he was also to provide companionship and support for Pie Neveu, an Assumptionist who had functioned in Russia since 1906 and had been secretly ordained a bishop in 1926. Fr. Braun was the first in a series of American Assumptionists to take up duties as the lone Catholic priest ministering to American and other Catholics in Moscow.[21]

Fr. Braun had been the first postulant to enter the Assumptionist novitiate in Bergerville in the Province of Quebec. He and some of those who followed him had at long last begun the process of staffing the college with American priests. As their numbers were growing, European Assumptionists were also being called home. As long as the college faced both high debt and low enrollments in a time of economic recession, there could be no serious attempts at expansion. Nonetheless, visions of future possibilities continued to inspire the leaders of the Congregation and the college. One of the oldest ideas, an American alumnate, came back to life during Fr. Armanet's tenure as superior and vicar provincial.

Since 1915, students interested in becoming priests had been brought together regularly as members of the *Cercle St.-Jean* (St. John Circle). A priest had been assigned to serve as their leader and to help them develop their piety and their understanding of the priesthood. Although the college produced many future priests, the belief persisted that a separate institution would be a more effective way of promoting vocations, especially Assumptionist vocations. In 1932, Fr. Armanet proposed to the Bishop of Springfield, Thomas M. O'Leary, that the shrine and the French-speaking parish of St. Anne in Fiskdale, Massachusetts, be entrusted to the Assumptionists, who had engaged in weekend ministry there for many years. He said he would assign a full-time pastor, who would be assisted on weekends by three or four other priests who could preach and hear confessions in several languages, to the benefit of the diverse groups of pilgrims who visited the shrine. He also proposed a house for closed retreats and the opening of an alumnate or apostolic school for young men attracted to the priesthood. The bishop was positive in his response, but urged secrecy while he considered the matter.[22]

Fr. Armanet encountered resistance on the issue of the apostolic school from his provincial superior in Paris — and his predecessor in Worcester — Fr. Clodoald Sérieix. Fr. Armanet wrote to his superior general in Rome, Fr. Gervais Quénard, that Fr. Sérieix "has always been opposed to this project. He thinks the College is sufficient to procure recruits for us [the Assumptionists], but no one here shares his opinion." From New York, Fr. Adrien Buisson pressed Fr. Sérieix to change his mind: "If an apostolic school could be founded, recruitment would be easier. All the savings of our [parish] works in New York go to the novitiate, which has hardly any novices; an alumnate would furnish, in time, ten or more each year."[23]

By early 1934, two years after making his proposal for Fiskdale, Fr. Armanet still had not received an answer from the bishop. He began asking for permission to buy a house elsewhere for the apostolic school, provided he could raise the funds without adding to the debt of the college. "Many beautiful properties," he wrote, "are for sale for *almost nothing*." When he met again with Bishop O'Leary in April 1934, the conversation took a somewhat ominous turn, perhaps revealing why the bishop had been concerned about keeping the proposition secret. "The questions the bishop asked me show that the *French* college still frightens him," Fr. Armanet reported, "because we are said (wrongly) to favor [French-] Canadian nationalism. I answered him that our students are American to the marrow and that if we do give them a French Catholic culture, we also produce good and loyal citizens. 'Yes, give them a French culture, but not Canadian,' the bishop responded. He seemed satisfied by my explanations and added, 'personally I am in favor of giving you what you want, but I must consult my council.'" Five months later the diocesan consultors unanimously told the bishop that entrusting Fiskdale to the Assumptionists was "inadvisable."[24] As events unfolded, the Assumptionists would not be entrusted with the parish and shrine of St. Anne until 1955.

Despite his council's negative recommendation regarding Fiskdale, Bishop O'Leary did grant the Assumptionists permission to open what Fr. Armanet called a "recruitment house" in his diocese. By early 1935, both the provincial and general chapters of the Assumptionist Congregation had approved the creation of an apostolic school and the admission, the following September, of poor boys wishing to become Assumptionists. Initially, they were to be housed and taught at a reduced fee at Assumption College, but within a year they were to be transferred to a new location.[25]

Fr. Armanet regretted that his Fiskdale plan had failed, "because the offerings of the pilgrims would by themselves have assured the subsistence of the students," but he continued to look at properties and sought permission from his Assumptionist superiors to buy an estate in nearby Barre, Massachusetts.[26] Though no action was taken on that matter, there were fifteen students admitted to the apostolic school at the college in September 1935. The number had "melted" to nine by March 1936.[27] By December 1936, the apostolic school was in its second year, still at the college, where, as things turned out, it would remain for lack of funds and staff to support a separate establishment. A pledge drive among the Franco-American clergy had helped to support the first boys,[28] but by the end of 1936 the new superior, Fr. Rodolphe Martel, was noting that they had become a part of the college's financial problems: "Excuse me for not yet being able to respond to your requests," he wrote to Paris. "Our treasury is still dry. And that shows you that this year is a little worse than the preceding one. There are fewer donations, students who pay less and for whom it was necessary to make dis-

Fr. Odilon Dubois and members of the Apostolic School, 1938–39

counts, many repairs necessary to prevent the building from going to ruin, not counting the students of the apostolic school, of whom there are 19."[29] As an alumnate within the college, the apostolic school would serve in the upcoming years as an important source of priestly vocations for the Assumptionists, for other religious orders, and for the diocesan clergy.

Change also visited the other community of religious on the Greendale campus. Since 1915, cooking and other domestic services at the college had been entrusted to the community of nuns founded in Greendale by Fr. Marie-Clement Staub, the Sisters of Joan of Arc — or Sisters of St. Joan of Arc as they were known following Joan's canonization in 1920. As vicar provincial for North America (the representative of the provincial superior in Paris), Fr. Armanet objected to some of the ideas and behavior of Fr. Staub, who continued to serve as spiritual director of the nuns at their mother house in Bergerville, Quebec, adjacent to the Assumptionist novitiate. Among other offenses, Fr. Staub was perceived as competing with the college in his fund-raising activities. As superior of the college, Fr. Armanet found that both Fr. Staub and the mother superior from Bergerville repeatedly inserted themselves into the Greendale picture, making demands that were difficult to deal with. In March 1931, he raised with his provincial superior the possibility of bringing in another religious community to replace the Sisters of St. Joan of Arc. Over Fr. Staub's strenuous objections, the Antonian Sisters of Mary were invited from Quebec; they accepted and took up their duties on July 17, 1932.[30]

Worcester Benefactors

The college's financial situation led Fr. Armanet into what some considered an unlikely course of transactions with two prominent Worcester citizens, John M. Tinsley, English-speaking vice president of Crompton and Knowles Loom Works, and Mrs. Homer Gage, daughter of the Knowles who had cofounded the loom works. Mabel Knowles Gage was a zealous Francophile, a philanthropist, a Protestant, and the wife of one of Worcester's most respected industrialists. Homer Gage, the president of Crompton and Knowles, was also one of the city's most respected surgeons.[31] John M. Tinsley, who acted as a financial advisor to Fr. Armanet, coveted membership in the French Legion of Honor, a distinction Fr. Armanet sought for him, and finally secured in 1931 through the French ambassador, Paul Claudel.[32]

The Worcester businessman suggested that Fr. Armanet discreetly approach Mrs. Gage, an officer of the Legion of Honor, about the possibility of financial support for the college. That strategy bore fruit in 1934, when Fr. Armanet himself received the Legion cross at the June 13 graduation exercises. "This distinction accorded me by the French government," he reported, "was not unrelated to a gift Mrs. Gage just gave us.... [T]hat very evening after graduation, while dining at her home with the consul general of France I obtained with no difficulty [a promise] that she would pay for our new driveway. She donated $7,573 to that effect and believe me it is beautiful." Access to the college from West Boylston street had for decades been available only via the badly maintained city street known as Assumption Avenue. From 1934 on, arrivals and departures were made via the paved 400-foot-long, tree-lined driveway and sidewalk known as Mrs. Homer Gage Avenue.[33]

Through Ambassador Claudel, Fr. Armanet also obtained a subvention from the government of France, which brought in a total of $1,000 in 1930 and 1931 and another $800 in 1932.[34] With Tinsley's help, he searched in 1932 and 1933 for an American institution that would lend the college money to pay down its debt to the principal lending agency, the Artisans Canadiens-Français of Montreal, which was to be paid in Canadian dollars. Since the Canadian dollar had fallen 10 percent in value against the American dollar, American money deposited in Canada would suddenly gain 10 percent in face value before being applied to the loan. The effort failed when Fr. Armanet was "not able to find money anywhere."[35]

Pleas from Europe

Much of the superior's time was devoted to fending off requests from European Assumptionists for loans or gifts, especially requests from his provincial superior, Fr. Clodoald Sérieix, the man he blamed for the college's large debt and for the gifts and loans to Nîmes. To the superior general, Fr. Gervais Quénard, Fr. Armanet insisted in 1930, "when one has a large debt one must not burden oneself further to help others. Know that when we can realize some savings we will be happy to have our dear France benefit from it." In 1932, he listed for Fr. Sérieix the economic and financial constraints the college was under, including the loss of nearly $8,000 deposited in the failed Bancroft Bank in Worcester. "I give you these details so that you might well understand the difficulties in which we find ourselves and that promise to get worse tomorrow. You will therefore not accuse me of bad will if I do not seem enthusiastic about the idea of lending you $40,000. Where would I find it?" He sent a message to be published in the Assumptionist newsletter, explaining that "for the first time in the twenty years I have been in the United States [paid] Mass intentions are lacking. Not only can we no longer afford to furnish them to our brothers in Europe, but we do not have enough for ourselves and for the fathers of the novitiate in Quebec. So it is useless to write us asking for some."[36]

Despite the dire circumstances, the college did contribute money to the province[37] and paid the annual provincial dues and taxes required by a new financial system established around 1934. The cost to the college of those charges in 1935 was $4,025.[38] The largest single contribution to the provincial superior in Paris was $20,000, recorded when the college in 1933 turned over to Fr. Sérieix the $20,000 once loaned by him as college superior to the General House and transferred — with his permission — to the college in Nîmes. Fr. Armanet and his councilors in Worcester had long given up on collecting that debt, and they were more than willing to let Fr. Sérieix deal with Nîmes while the college was credited with a $20,000 contribution to the province.[39]

Perhaps the single most important achievement of Fr. Armanet's financial stewardship was the one described in his April 1935 observation: "During the last five years, and despite the economic crisis, we have not only paid our interest very regularly, but we have also amortized by $30,000," leaving a debt of $265,000.[40] In 1935, he received the unexpected news that the General House had come into cash which it was willing to lend to other houses at a low interest rate to help them reduce their debts. Fr. Armanet immediately accepted the $70,000 offered at 3 percent interest and applied the whole amount to the debt, on which the college was paying 5 percent. The result of the reduced principal was an annual savings of over $1,300 in interest.[41]

Fr. Armanet was skilled at identifying benefactors, such as Mrs. Gage, who would respond to requests to finance specific proj-

Statue of St. Joan of Arc

ects, like the entrance road, that would add to the ambience of the college. In 1931, a priest from Fiskdale underwrote the creation of the college's first flag, on which appeared the college's first motto: *Je maintiendrai* (I will uphold). Fr. Armanet explained its full meaning as, "I will uphold the faith, the language and the traditions of my Fathers."[42] He also secured funding for a grotto of the Virgin Mary to replace the one originally built in 1909, which had been demolished to make room for construction in 1926.[43] In 1933 a priest friend from Millbury paid for the construction of the new grotto in a conspicuous location that would prove attractive to visitors and pilgrims.[44] In 1934, at the expense of other friends of the superior and the college, a bronze statue of Joan of Arc and a granite pedestal to display it were commissioned and installed at the summit of Mrs. Homer Gage Avenue.[45]

Preparing the Succession

In the spring of 1935 Fr. Armanet was anticipating the conclusion of his second three-year term as superior and president. He welcomed the prospect of giving up the duties of that office while continuing to serve as vicar provincial for North America. It was a position that in his opinion really required more time than he had been able to devote to it. The work called for a considerable amount of travel, he believed, which was not possible for the college superior, "riveted as he is to his post." He recommended that his subprior in Worcester, Fr. Rodolphe Martel, be named his successor at the college. He testified that the 55-year-old native of France had demonstrated "authority, know-how, energy, plus a brilliant intellect." The only drawback, Fr. Armanet wrote, was Fr. Martel's inability to speak English. Fr. Armanet, who had served ten years in England before coming to the United States in 1912, suggested that Fr. Martel's not knowing how to speak English would not pose an insuperable obstacle, "as long as the [vicar] provincial knows it." The problem would also be extenuated, he argued, by the fact that the college superior could always call upon the help of English-speaking subordinates "in certain circumstances in which English is necessary."[46]

Fr. Rodolphe Martel, A.A.

Fr. Armanet and his two predecessors, Frs. Marie-Louis Deydier and Clodoald Sérieix, had all spent years in England and had mastered the language before coming to the United States. Why had Fr. Martel, who had been in Worcester for fifteen years, still not learned to speak English? "For lack of opportunity to speak it," he wrote to his new provincial superior in Paris, Fr. Bernardin Bal-Fontaine. Fr. Martel wrote that he could read and understand the language well, but he was not able to give a public address in English. On the other hand, he believed, there were "not many circumstances in which that has to be done."

Though he had not yet been appointed superior, Fr. Martel took the liberty of making recommendations about how the staff might be reorganized should he be named to the post. His first suggestion was a striking one: that the second-ranking position in the local community, the post of subprior, might go to a twenty-seven-year-old Franco-American, Fr. Wilfrid J. Dufault. Fr. Dufault was a 1929 graduate of the college who had been ordained in 1934 and had just spent the 1934–35 academic year teaching philosophy at the college. In Fr. Martel's opinion, Fr. Dufault was "serious, [knew] English well, and [could] be trusted." On the other hand, he was "still young and inexperienced, and his appointment might not be accepted warmly." Among the more seasoned members of the staff qualified to become subprior, Fr. Martel proposed 49-year-old Fr. Odilon Dubois, who had been serving as a teacher and administrator in Greendale since 1919. He was "a veteran, very serious, [and] pleasant in conversation." As he thought about the generational difference within the community, Fr. Martel seemed to warm to the prospect of the older man as second-in-command. Fr. Dubois, he remarked, "almost never leaves the house, [and] he has a perfect knowledge of all the traditions. He is probably the best professor we have, and the young ones have a great deal to learn from him."[47] On August 15, 1935, Fr. Martel succeeded to the superiorate and the presidency of the college and Fr. Dubois became subprior.

Since 1923, the office of vicar provincial had been held by the man who was simultaneously superior at Assumption College, first Fr. Clodoald Sérieix and then Fr. Crescent Armanet. Now Fr. Armanet was moving back to New York City, but, as he had hoped, he was taking with him his office of vicar provincial, an arrangement that placed the new head of the college, Fr. Martel, under the supervision of his predecessor, Fr. Armanet. This was a return to the structure of 1919–23, when Fr. Marie-Louis Deydier as superior in Worcester had found many an occasion for complaint about what he regarded as interference by his predecessor, Fr. Tranquille Pesse, who had left Worcester to serve in New York as common superior for North America (the title used before the establishment of provinces in 1923).

The restored arrangement would lead to some friction between college president Martel and vicar provincial Armanet as well. One dimension of the problem was addressed in a 1937 letter from Fr. Martel to the man who supervised both of them, the provincial superior in Paris. "Is it Fr. Crescent who is dealing with you about our [Worcester] business?" Fr. Martel demanded of the provincial. "In that case, it's useless for me to write to you. Or else his governmental authority [over the college] is null, and then it's useless for me to deal with him. Dealing with two at the same time doesn't work."[48] As in the troubled days of Fr. Deydier, friction also developed when the vicar provincial in New York sided with some of the Worcester religious in opposition to their local superior.

Over the course of the next decade, as Fr. Wilfrid Dufault and the young Franco-American Assumptionists began to develop visions for the college — visions not entirely shared by Fr. Martel — they would sometimes find a potent ally in New York.

Overcoming the Paralyzing Debt

As he began his service as superior during the Depression summer of 1935, Fr. Martel, like Fr. Armanet before him, stared with dismay at the college's debt and could not forget the $20,000 loan Fr. Sérieix had made to the Assumptionist General House and had then allowed to be transferred to the college at Nîmes. In Worcester, financial progress had been made, Fr. Martel allowed, "but our debt is still enormous: 249,000 dollars. That paralyzes a work and a superior. I hope that such borrowing will never happen again. If only the [$]20,000 we loaned to the General House had been returned and not stolen along the way." On another occasion he wrote, "If there is anyone who has a horror of debt, it's me. . . . Our work will be paralyzed for at least twenty years." Not long after he added, "Each day I have occasion to tremble at the follies that have been committed. I told Fr. Clodoald [Sérieix] that one day his successors would damn him. I did not think that the successor would be me."[49]

Even more than his predecessor, Fr. Martel spent many hours explaining to his European superiors, especially to the provincial superior in Paris, why the college could not respond fully to repeated requests for financial assistance. In June 1936, he sent a check for $1,500 to cover the annual dues the province had imposed ($40 per priest, $20 per brother, and 10 percent of mass stipends), pointedly remarking, "It is to be noted that our dues are 3 times greater than those paid by the religious in France." The provincial superior had asked additionally for a "gift" amounting to 10 percent of the salaries the religious would be making if they were being paid.

Fr. Martel asked for a delay until he could come up with the money. "At no price will I take out a loan," he wrote. "The one on our shoulders is heavy enough." He recalled that in addition to the $20,000 college loan shifted from the General House to Nîmes — the loan Fr. Crescent Armanet had give up on and turned over to Paris to try collecting — there had been $20,000 loaned directly by the college to Nîmes. Very little of that had been paid back, and "I had counted on [repayments] from Nîmes for [the money to cover] these gifts, but it seems the gentlemen over there are close-fisted. We were less hesitant when it came to providing them with the money for their new college; and besides that[,] they profited from the [$]20,000 that were stolen from us. But I don't wish to recriminate, we will do what we can; furthermore, it is absolutely necessary to get out of the state of stagnation in which Fr. Clodoald [Sérieix] mired us by his grotesque ideas."[50]

When that explanation also failed to stifle calls for financial assistance, Fr. Martel declared, "We have served the province as well as any house and probably much better than most. One must estimate what we have given over the last 10 years at least 7 1/2 million francs [around $500,000]. And that despite the foolish debt that had to be carried."[51] Three months later he searched for new ways to make the same points: "*We are not refusing to help you,*" he underlined. Nîmes had reneged on a promised payment of between $1,000 and $2,000, "and since then we have not received anything we could take out of our treasury. Today we have exactly [$]2807.87. If we were to pay the debts due up to today we would need at least 3,000 dollars. In fifteen days we would need [$]3,000 more. So we live from day to day. . . . *Our financial situation is not without danger,*" he further underlined. "People are perhaps tempted to consider America as an inexhaustible gold mine. That's false. . . . Here we will need 7 to 8 good years to free ourselves from the vise that is squeezing us.[52]

Despite his sense of financial paralysis, Fr. Martel in the early months of 1936 was hoping to come up with the funds for a respectable student library. "We have nothing to show outsiders who visit us," he wrote. "True, it's not for them that one has a library, but we can't show it to them because we do not have it." A few months later he was able to report that "Mrs. Gage just gave us 1500 dollars to start the students' library." During the summer of 1936, Frs. Wilfrid Dufault and Edouard Moreau, along with long-time lay professor Valmore X. Gaucher, transformed the day students' study hall into a library, carting in books and cataloging them according to the Dewey Decimal System, "to give Assumption a library established on the same plan as those of the largest institutions in the country," the college magazine reported. "It's difficult to imagine a college without an appropriate library," remarked the superior and president, Fr. Martel. "Ours will total around ten thousand volumes in a short time."[53]

Another early ambition of Fr. Martel was to fulfill a promise administrators had been making to the students for a number of years. Of the three sports in which Assumption students competed at the intercollegiate level, the tennis team had flourished on twelve courts built in 1924, and the basketball team practiced and played in the modern gymnasium completed in 1927. What was eagerly awaited was a high-quality baseball field for general student use and to accommodate the exploits of the popular college baseball team. The gift of a field came about after Fr. L.P. Rodier, pastor of Holy Name of Jesus parish in Worcester, was the principal speaker at the annual superior's Feast Day banquet in 1936. He celebrated the spirit of the Assumptionists in nineteenth-century France, "their spirit of struggle, of combat, of attack in defense of the church, of the papacy, of the undeniable rights of evangelical truth and Christian morality in that troubled century." In Worcester, he remarked,"our nuns, those valiant humble women who have devoted themselves to our parochial schools, have saved our parishes. Your college at a higher level of education has saved the Franco-American race by providing it with an intellectual elite, with educated and enlightened leaders." When Fr. Rodier later expressed a desire to contribute something "visible and enduring" to the college, Fr. Martel apparently had a suggestion. In 1937, Fr. Rodier donated $5,000, later increased to $8,000, for the construction of what became Rodier Field.

In 1937, the college was reminded once again of the steps it would have to take if it was to receive approval from an accrediting agency. After a representative of the National Catholic Educational Association had visited and made his report, the college was refused accreditation on a number of serious grounds: the use of the same laboratories by the high school and the college; faculty members' lack of graduate training in the fields they were teaching; and the inadequacy of both the library and the laboratories. To this, Fr. Martel's principal response was to suggest procedures for making sure future Assumptionists with teaching potential finished their undergraduate degrees at the college before going off to the novitiate in Bergerville.[54] While this would have helped to solve the old problem of a widespread lack of degrees of any sort among the teaching corps, it did not respond to the more recently perceived need for more graduate degrees among the faculty.

While Fr. Martel was securing outside funds for specific improvements, he laid siege to the debt that troubled him so greatly. A year after his appointment, he was able to report that the college had reduced by $20,000 what it owed to the finance company in Montreal, bringing that part of the debt down to $170,000. The college also owed to the Congregation the $70,000 sent over in 1935 to reduce interest costs, and to the architect of the 1926–27 construction it still owed $4,000, on which the college regularly paid interest.[55] Student enrollments began to recover in September 1935, rising from 197 the previous year to 242, due in part to the entry of 19 students into the apostolic school.[56] For the next six years, the size of the student body was comparable to what it had been in the middle 1920s, averaging around 250 students, two-thirds of whom were in the high school.[57] American entry into World War II at the end of 1941 led to higher wages and a surge in enrollments in the high-school classes.[58] The resulting income provided even more ammunition for Fr. Martel's war on the debt, and in January 1945 he could announce proudly to his staff that the Montreal portion of the debt had been completely repaid. The $70,000 owed to the general house remained outstanding.[59] Along the way the aggravation over the accounts with Nîmes had been finally put to rest. On October 18, 1944, "given the situation of Nîmes which has been strongly tested by the war" the house council "decided to forgive them the debt of around $15,000 that they have yet to pay on the [original] loan of $20,000 made by Worcester several years ago."[60]

War

Although the boom in high school admissions had brought in much-needed revenue, it was accompanied by a dramatic decline at the collegiate level. Assumption College would almost surely have perished during World War II had it not functioned as a high school as well as a college. Indeed, as older students volunteered for military service or were conscripted, the college-level program virtually disappeared. A few months before Pearl Harbor, there were 81 students studying for their bachelor's degrees; nine months after Pearl Harbor, there were 46. In March 1943 there were 20. Meanwhile, the high-school enrollments kept expanding rapidly. In September 1944, in a student body of 349, there were 9 college students, supplemented by 12 young Assumptionist choir brothers (candidates for the priesthood), many of whom had interrupted their college careers to enter the novitiate in Quebec and had then returned as draft-exempt religious to pursue their studies.[61]

Gaston Henry-Haye, Vichy Ambassador

Before the United States entered the war, the leaders of the college had to confront the emotional and political implications of the fall of France to German forces in June 1940 and the post-armistice division of the nation into a northern zone occupied by the Germans and a southern zone, with its capital at Vichy, under the authoritarian governance of Marshall Henri-Philippe Petain. As longtime antagonists of the Third Republic, some Assumptionists were quick to blame the vices of liberalism and anti-Catholicism for the military collapse and to find in Marshall Petain hope for a revived Catholic France. On December 6, 1940, the college warmly welcomed a visit from the Vichy government's ambassador to the United States, Gaston Henry-Haye.

Fr. Henri Moquin, A.A.

Among French-speaking New Englanders, the Assumptionists were hardly alone in their support for the Vichy government. The president of L'Union St. Jean-Baptiste, the largest Franco-American fraternal organization in New England, had suggested the ambassador's visit to the region and accompanied him. At a dinner hosted by the college, members of the local clergy were present, along with representatives of other Franco-American societies, and individual Franco-Americans from the city. The occasion was also graced with the presence of the philanthropic Mrs. Homer Gage and Mr. John Tinsley. As head of the college Fr. Martel conveyed "his grateful admiration for the magnificent work of purification and rectification Marshall Petain is pursuing with typically military vigor," emphasizing "in particular the repression of all antireligious legislation, especially that against religious congregations." "I drink," he concluded, "to the resurrection of France."[62]

When Assumption men went off to war, Fr. Henri Moquin launched a mimeographed newsletter, "Assumption at War," that did its best to reach the hundreds of former students in uniform and to keep them informed of one another's activities.[63] While that link was being successfully maintained, communication with Assumptionist authorities in Europe became impossible between 1942 and late 1944.[64]

Pleading for an American Superior

About a year before the long silence began, one of the young Franco-American Assumptionists managed to get a letter into Vichy France, addressed to Fr. Felicien Vandenkoornhuyse, the first assistant to the superior general, at the Assumptionist house in Nîmes. The second — and normally the final — three-year term of Fr. Rodolphe Martel as superior in Worcester was coming to an end. The Assumptionists had received from the Vatican permission to keep the superiors of their many houses in place until the end of the war,[65] but Fr. Antonio Laberge was appealing to the highest authorities in the Congregation to make a change in Worcester, despite the international crisis.

The work being carried on in Worcester was "gravely compromised," he reported, because the critical task of securing accreditation had been and was still being neglected. Boston University, which had been "the sole refuge in New England of our future doctors," had just refused admission to the college's best graduating senior, and another member of the class had for a time been unable to get into a school of naval aviation because the college was not accredited.

Fr. Laberge traced this "neglect" all the way back to 1917, the year the college received a charter authorizing it to grant degrees but did not follow up immediately by seeking accreditation from an association of colleges and universities, an effort he believed would have "easily" succeeded (a dubious judgment, given the lack of degrees among faculty members). Fr. Laberge could attribute the failure to seek accreditation only to "ignorance of the functioning of the American university system, altogether different from the French organism." According to Fr. Laberge, in order to justify the college's exclusion from the ranks of accredited institutions, some of the French priests indulged in the belief that the education they offered was superior to that available in American colleges, from whom they had nothing to learn about educational methods. "Unfortunately," he added, "those of us who have attended American university centers now know that our pretended superiority is in large part illusory, and that we would perhaps have some interest in knowing certain American methods that are even being utilized in France. But there it is: the truth is that we are behind France as well as America."

Fr. Laberge then moved on to the "delicate subject" at the heart of his letter, a subject about which he and several of his Franco-American colleagues were in agreement: Since getting accredited was a matter of "vital importance" for the college, and since under the leadership of Fr. Martel there was little hope that the necessary steps would be taken, and since leadership of the college would in time devolve upon the Franco-Americans, "we venture to express to you our desire soon to have a Franco-American superior to lead us." This view, he wrote, was shared by Frs. Armand Desautels, Wilfrid Dufault, Ulric Charpentier, and Henri Moquin, though the letter itself was solely his responsibility.

Fr. Laberge stressed that the request was not in any way inspired by antipathy toward the French fathers, though he believed that among them only one currently at the college had an accurate grasp of the institution's needs, and that individual would not be suitable as superior and president because he did not speak English fluently. Fr. Laberge added another consideration: that the support of the Franco-American clergy and of the alumni would be strengthened by the appointment of a Franco-American superior. Having gone so far already, Fr. Laberge was willing "to designate the one who would rally the greatest support. . . . The Franco-American individual who seems to us most suited to be successful is Fr. Henri [Moquin], who has given proof his capacity."[66]

Fr. Moquin had been ordained in 1933 and thus had, along with Fr. Ulric Charpentier, the most years in the priesthood among the individuals for whom Fr. Laberge claimed to be speaking.[67] Whether Fr. Moquin's seniority was a consideration in his being designated as the preferred candidate is not evident. Fr. Laberge pointed to Fr. Moquin's success as prefect of studies, his popularity among clergy and alumni, his ability to get things done despite obstacles placed in his path, and the likelihood that all members of the religious community could embrace him warmly as superior. Whatever the reasons for Fr. Moquin's private selection by his colleagues, the Assumptionist authorities in Europe did not grant the extraordinary request for a transition to a Franco-American superior in 1941. Like so many other matters, this one would have to await the return of peace. Though it did not produce the desired effect, Fr. Laberge's letter documented a growing sense of distinctiveness among the Franco-American Assumptionists as well as an intensifying eagerness to assume responsibility for the college whose future was so intimately bound up with their own.

The Vanishing Franco-Americans

Their future and that of the college were also intimately bound up with the future of the Franco-American population of New England, a community that was becoming increasingly Americanized. A growing proportion of the population was abandoning the use of French in the home and sending its children to public high schools rather than to French-speaking parochial schools. To adapt to that situation, the college took a significant though not entirely unprecedented step in 1941. From its earliest years, Assumption had, when it seemed desirable, provided "preparatory" classes for boys entering the high school with too weak a command of French to be able to pursue the normal curriculum immediately. As recently as 1934, in an attempt to attract new students, Fr. Crescent Armanet had announced the inauguration of such courses, though he had later been forced to announce their cancellation due to insufficient enrollment. In 1941, the college for the first time announced that it would accept into the high school young Franco-Americans who did not know French at all. They would be introduced to the French language and begin Latin in English, but by their third year they were to be ready to rejoin their classmates who, from the beginning, had studied both Latin and French in French-speaking classes.[68]

Fr. Martel anticipated the inevitable complaint that the French character of the college was being diluted. He asked whether it was more desirable to "abandon" Franco-American boys who did not know French, and thus push them toward a purely English education. He answered himself: No one among the Franco-Americans "could say anything but an energetic No."

"That is the entire reason for the small change we intend to make in our program," he tried to reassure skeptics. "The old program subsists.... There is no need for anyone to get alarmed; there is reason for everyone to rejoice." He reminded the friends of the college that in earlier years the faculty would not have been capable of undertaking such an effort among English-speaking boys. "This change has been made possible thanks to the presence in the religious professorial staff of young Franco-American Assump-tionists full of zeal, of learning, and of virtue. They know the college, its strengths and its weaknesses, having been students of this college themselves. They have a great deal of ambition for it. They are counting on the friendship of their old comrades. They have decided to go to work. And they dream only of advancing the kingdom of God."[69]

The discomfort with which Fr. Martel addressed the matter of admitting boys who did not read or speak French was reflected in other thinking about the future of a French-speaking Franco-American population in New England. In 1940, Fr. Martel raised the question of the group's potential for survival in the face of the decreasing use of French in an English-speaking country, of intermarriage between members of different ethnic and language groups, and of the "sad but sure desertion of Franco-American churches." He called for a moral commitment to resist those tendencies through education."It's up to you, parents, to push your children to become leaders; up to you, young children, to want to become leaders."[70]

Sometimes even positive news produced, upon reflection, reasons for anxiety or complaint. An annual scholarship examination, organized for prospective high-school freshmen beginning in 1942, attracted over 230 hopeful competitors the first year, and over 300 the next, leading to the inference that "a college of over 1200 students of Franco-American origin could easily be created. There would have to be more concern in families and in the leading centers of higher education. On this terrain, they [Franco-Americans] are letting themselves be beaten by others. It's a tactical error, and it will be necessary to pay for all sorts of consequences that will result."[71]

Though they were trying everything they could to insure the college's future, the Assumptionists depended upon decisions that were being made day in and day out by individuals and families of the population they sought to serve. Because they regarded those decisions as morally laden, spokesmen for the college sometimes adopted a scolding tone as they carried to Franco-Americans the message that support for the college was a duty not to be neglected. Sometimes they went so far as to warn that if the college were to fail, it would be the Franco-American population, not the Assump-tionists, who would have to accept the blame. Speaking at a parochial high school graduation ceremony in 1943 one of the Franco-Americans, Fr. Antonio Laberge, cautioned that all the "money and the sweat of your fathers" would have been expended in vain if the current generation of Franco-Americans did not take up their share of the cultural burden.

"Let us in all sincerity, me as well as you, examine our consciences. Yes, I am indeed saying: me as well as you, because I would be as guilty as you if I did nothing to remedy this state of affairs. I would even be more guilty than you, because as a member of the faculty of the only Franco-American college in the United States, I have a grave obligation to strive with all my might for the intellectual and moral welfare of our element."

Having personally identified with his listeners, Fr. Laberge hastened to relieve the Assumptionists of any laxity in the battle for *la survivance:* "It was particularly for the Franco-Americans that the Assumptionists came to New England. WE HAVE NEVER FAILED IN OUR MISSION: and I can affirm to you today, without fear of contradiction, that IF EVER THE ASSUMPTIONISTS CEASE TO CARRY ON A FRANCO-AMERICAN WORK, IT WILL ONLY BE BECAUSE THE FRANCO-AMERICANS THEMSELVES WILL HAVE MADE THE TASK IMPOSSIBLE FOR THEM."[72]

A year later, Fr. Laberge called for support of the "new Assumption College" that would arise following the end of the war and the return of college-level students. "That is where bilingualism... will have, so-to-speak, its fortress on American soil."[73]

Preparing for a "New College"

The "new Assumption College" was the centerpiece of a fund-raising campaign launched by the superior, Fr. Martel, in October 1944. He invited Franco-American leaders to a meeting scheduled for November 12 to consider the college's building needs. On November 3, a preliminary meeting was held to plan a drive "whose funds will be destined for the urgent construction of a new college on Burncoat Street," at the eastern edge of the college campus. Supported by many Franco-American societies, and especially by L'Union St. Jean-Baptiste, the campaign set a goal of $250,000 to help provide new facilities in anticipation of the return of college students after the war, with the "double advantage of giving space to the high school students... and of separating the college from the high school; which would give the college a situation that would conform more to that of analogous institutions in the country."[74] A circular issued for the campaign featured an architect's sketch of a ten-building "new college" facing Burncoat Street on land that, coincidentally, had been purchased with proceeds from the first major fund drive, at the end of World War I. The new college was to stand back-to-back with the older facilities that faced West Boylston Street, with a large expanse of playing fields separating the two. The brochure was

Architect's sketch of "new college" on Burncoat Street

headlined, "For Assumption College in Worcester and French survival [*survivance*] in New England."[75]

The projected campus on Burncoat Street was to become a controversial matter within the Assumptionist community, in part because the new location was itself a product of disagreements about what to do about the college once the war was over. The Franco-Americans, among whom Frs. Wilfrid Dufault, Armand Desautels, and Henri Moquin had emerged as leaders, were convinced that the college urgently needed accreditation if it was to survive, and that it would not succeed in that quest before the college was physically separated from the high school. Some of the European-born Assumptionists were of the same mind, including the vicar provincial, Fr. Crescent Armanet, based in New York. Not included in that consensus was the superior of the college, Fr. Rodolphe Martel, and at least two other French-born religious.

Moving the college proper from Greendale had become an active project early in 1943. At that time, the college's frequent benefactor, John F. Tinsley, was serving as the executor of the estate of Theodore T. Ellis, former publisher of the *Worcester Telegram.* He visited Fr. Martel to suggest that the Assumptionists might be interested in purchasing the buildings and approximately one-hundred-fifteen acres of land at 245 Salisbury Street, on Worcester's West Side that were part of the Ellis legacy. As Fr. Martel recalled in 1946 with some humor, "I replied that I was not interested and that besides we did not have money to spend. Some people came to take me by force. As a matter of conscience I wrote about it to Fr. Crescent [Armanet], who came at a full gallop to visit it, found it to his taste, ran to the bishop for permission to make the purchase, but didn't get it. The bishop later had us informed that he wanted to see us develop in Greendale."[76]

Fr. Armanet had indeed "run to the bishop" (Thomas M. O'Leary, Bishop of Springfield) on March 1, 1943, and had been given no definite answer, although the bishop had let him know that he had some interest in the Ellis property himself, for an undisclosed diocesan purpose. In June, Fr. Armanet, after failing in his efforts to see the bishop a second time, had sent Fr. Armand Desautels from the college to Springfield to ask if the bishop had come to a decision. It was at this meeting that Bishop O'Leary said he "absolutely did not want" the Assumptionists established anywhere in his diocese but in Greendale, where, he said, they had enough land to build upon. Urged on by the younger priests at the college, Fr. Armanet had then set off on a series of visits to bishops in other New England dioceses, but he was unable to convince any of them to welcome the unique Franco-American college into his jurisdiction.[77]

As superior in Greendale, Fr. Martel had not grown any fonder of what he regarded as the vicar provincial's interference in the business of the college, and he considered Fr. Armanet's series of failures in chanceries across the region an embarrassment to the Assumptionists. "I said to myself: We are really looking too stupid," he later wrote. "We have to do something that will be successful. During the summer of 1944 I thought the matter over again. I had our dedicated architect, Mr. [Albert] Roy, draw up a plan of a future college to be built on Burncoat Street. Everyone seemed enchanted by it."[78]

Although the fund drive would be successfully conducted, the "new college" was never to rise on Burncoat Street. Instead, Fr. Martel would return after the drive to an earlier idea. On March 27, 1944, he had asked the house council what the religious thought of plans drawn up by the architect for building on top of the gymnasium, which had always been intended as a platform over which a chapel and some dormitories would eventually rise. According to the council minutes, the councilors thought that, "given the fact that the current buildings could suffice again this year for receiving the anticipated students, it would be better to wait at least until next year before embarking on new debts ($100,000, not counting furnishings.)" Having received that negative response, Fr. Martel apparently did not further consult with the council on building plans or how such plans would be financed.[79]

By summer, the unpopular idea of expanding above the gymnasium seemed to have been shelved, A similar fate did not at that time befall the project many of the younger Assumptionists also considered unwise, the building a "new college" on Burncoat Street. To their dismay, Fr. Armanet, the vicar provincial they counted on as their ally, visited the campus in the autumn of 1944 and "suddenly declared the matter closed, and the new college would be built on Burncoat Street."[80]

So the fund drive was launched and, as Fr. Martel recalled, "the more success seemed assured, one would have said, the longer the faces of our young men seemed to get, some of them, anyway." Convinced as they were that accrediting agencies would balk at the college's continuing to share a campus with the high school, some of the younger Assumptionists may indeed have looked somewhat glum as they obediently fell into line. "We collaborated with the campaign as best we could," Fr. Dufault wrote two years later. "I distributed this circular as the others did, aligning myself fully with the opinions of the superiors." The members of the new generation put their shoulders to the wheel out of a sense of religious duty, but also because they agreed with Fr. Martel when he stated that "the perilous honor of forming the best men of our element is reserved for our college."[81] Knowing that the honor of forming an elite for the Franco-Americans would soon be theirs to carry, and recognizing the peril that honor might place them in, they strained to have their ideas respected as the postwar future drew near.

While the drive was being organized, the vicar provincial was able to reestablish written communication with the Assumptionist authorities in Europe after a break of more than two years. In a long report to Superior General Gervais Quénard, written on November 18, 1944, Fr. Armanet raised the issue of the long-anticipated Franco-American succession in Worcester. "It will soon be necessary to replace the superior, who is beginning his 4th term," he wrote. "The replacement, in the opinion of my council and of all the religious should be a Franco-American. No other, be he here [in New York City], be he in Worcester, could replace Fr. Rodolphe [Martel]. At the request of this last, I have discreetly consulted the religious to find out which one would best fill the position and *all* named Fr. Wilfrid Dufault. That is also very much my choice. He is a complete man: calm, poised, learned, strongly attached to his Congregation and a faithful observer of the Rule. The College has reached such a development that it is a matter of true necessity to give it a leader capable of measuring up to the presidents of American colleges and universities and able to speak their language well. That requires a person born in America."[82]

The fund drive was scheduled to be carried out in all of the Franco-American centers of New England between January 1 and March 18, 1945. The goal, as announced by Fr. Martel, was to raise at least $250,000 "for the construction of a new building for the college proper." Bad weather and other obstacles led the directors of the drive to extend the solicitation beyond March 18, and, by May, over $200,000 in cash had been collected. The campaign came to an official close on October 14, with $255,000 received and another $22,500 pledged. Solicitation continued, however, and by June 1946 proceeds would be approaching $300,000. However, the publicity associated with the drive had affected the college in an unanticipated way. Applications for admission to the high school were increasing dramatically.[83]

Although he expected to be replaced as superior during the summer of 1945, Fr. Martel made a momentous new choice that spring. He went back to the idea of building over the gymnasium, and this time he planned to use funds from the "new college" drive to pay for it. "Faced with this abundance of students," he wrote, "a decision had to be made; I called a meeting of my council, I presented the case; there were not two solutions, there was but one: it was necessary to finish the building over the gymnasium." Since the council minutes contain no record of such a meeting, it seems that Fr. Martel decided on his own to build a large chapel over the gymnasium, and above the chapel both a large dormitory to accommodate the sudden expansion in the high-school population and two additional halls intended for the library. He apparently agreed with the reasoning of the vicar provincial, Fr. Armanet, who believed that the college proper would "remain in a lethargic state as long as young men of 18 are being taken for the service of their country," while the high-school applicants were already at the door. During the fall of 1945 the new construction was begun,"but it made some people sick at heart," Fr. Martel reported, "to see that Fr. Rodolphe [Martel] could spend what he had succeeded in raising."[84]

There was indeed consternation over the fact that, on the eve of the Franco-American succession to the leadership of the college, steps were being taken that might restrict the options of the new generation for many years to come. Among the twenty-eight priests at Assumption College in 1945, only seven were from France. Virtually all parties expected Fr. Martel to be replaced by a Franco-American that spring or summer. In March 1945, the vicar provincial, Fr. Armanet, had been so sure the transition was imminent that he had asked his superiors in Europe "not to advise

Fr. Wilfrid yet of his nomination as Superior of the College as replacement for Fr. Rodolphe [Martel,]" who was directing the fund drive and "would be entirely disconcerted if he were replaced at this moment, and the 'drive' would suffer greatly." Fr. Armanet suggested that it would be "better to wait for the month of July" and asked that the major superiors "leave it up to me to set the exact date of Fr. Wilfrid's taking charge."[85]

By July, however, Fr. Martel was still in charge, as he would be for another year. The Assumptionist authorities in Europe decided that, before major personnel changes were made, chapter meetings should be held in each province or vicariate, followed by a General Chapter in Paris. On August 9 this policy was announced to the religious vacationing at Baker Lake by Fr. Martel himself. His ally, Fr. Odilon Dubois, noted in the "Archives de la maison," the journal he maintained on behalf of the community, "This news clipped the wings of quite a few canards." The vicar provincial, Fr. Armanet, was later told that, when Fr. Martel told the religious community in Worcester he would be staying on as their superior for another year, "the news was greeted... by a glacial silence." Despite the delay in changing superiors, the same message from Europe announced a significant first: a Franco-American, Fr. Wilfrid Dufault, was named subprior of the religious community and vice president of the college, replacing Fr. Odilon Dubois, who had held the post throughout the Martel superiorate and presidency.[86] Meanwhile the fund drive and the planning for expansion of the high school went on.

Explanations for what might have seemed to be the diversion of proceeds from the stated purpose of the fund drive appeared in various forms over the next months. Fr. Martel felt it necessary to point out, in the spring 1945 issue of the college magazine, the eventual benefit to the college of having so many well-trained high-school students from which to recruit collegians. He put into the mouth of a hypothetical reader a question about whether the use of drive proceeds for the high school would delay the plans for the college, to which he replied: "Not at all. We have anticipated everything and our resources will even be multiplied. At the propitious moment the college will rise up from the earth." A few months later, the same magazine was claiming that the decision to add to the high school buildings had been due to the college's "not being able to obtain the permissions and materials to start the new college." Elsewhere the magazine said, "It is understood that we will not neglect to build the new college as soon as that will be possible.... All the funds collected in the campaign will be used on that." The apparent financial strategy was to use — in a sense, to borrow — some or all of the proceeds of the fund drive, while planning to return the money for a college construction project at the appropriate time. At "another time in the not-too-distant future," Fr. Martel asserted, "we will succeed just as well [as this time], because the college that is to be built will not be built with $250,000. There will have to be several times that amount." At the end of 1945, Fr. Martel hailed the success of the drive, but emphasized that this was only a beginning. "The Franco-Americans are from this point on alerted to one of the important points of their problem; they have understood and seen, and they have chosen to continue and to accentuate their effort. The years to come will show what they can do."[87]

The anticipated arrival of many new students in September 1945 had triggered the decision to add to the high-school accommodations. However, when 415 students — a 20-percent increase — rolled up Mrs. Homer Gage Avenue on September 12, the new construction had yet to begin. Due to war-related delays in the arrival of building materials, construction did not begin until October. Exceeding 400 members for the first time, the student body included 30 individuals in the college proper and 368 in the high school. The fact that they were crammed into space that had previously been regarded as crowded for 350 students may have accounted to some degree for the fact that 21 of the high-school students left within the first month. Due in part to bad weather in December, the new facilities were still not usable for the second semester that began in January 1946.[88]

Moving from Worcester?

Meanwhile the steps required for the chapter meetings of the religious were under way, and, as they proceeded, the idea of moving the college to a new location suddenly gained new life. In October 1945, Fr. Wilfrid Dufault had written to the superior general, Fr. Gervais Quénard, with new evidence that the 45-acre campus in Greendale was simply too small for the two institutions to share it successfully. "Last summer," he later explained, "new reasons were learned that made this decision [to build on Burncoat Street] worrisome, for example, that 'Standards for Colleges' (Columbia, 1938) demanded a minimum of 60 acres for a small college etc. The growth in the number of students in the high school revealed how necessary it was to provide them with additional playing fields; finally it was ascertained that the initial installation on Burncoat Street would be much slower and actually more expensive than a more satisfactory installation elsewhere would be. That is why in October [1945] I submitted a report to Fr. General to bring him up to date with what we believed to be the pros and cons of a transfer of the college."[89]

On December 28, Frs. Wilfrid Dufault and Henri Moquin were elected to represent the religious community of the college in the vicarial chapter scheduled for early January 1946. According to Fr. Dufault, no discussions about reopening the search for a new location were carried on with Fr. Armanet, the vicar provincial, who had endorsed the decision to build on Burncoat Street. Yet, when Fr. Armanet arrived from New York to preside over the chapter meeting, he had already made up his mind, according to Fr. Dufault, to seek a place for the college in the archdiocese of Boston. The chapter voted unanimously to approve such an endeavor, but advised Fr. Armanet not to move on the matter until after the meeting of the General Chapter, scheduled for Paris during the summer. Fr. Martel, the college superior, who was a member of the vicarial chapter and had no enthusiasm for moving the college, later asserted that when the move to the Boston archdiocese came up in the chapter meeting, "all was decided without discussion."[90]

Fr. Joseph Pelletier, A.A.

In February 1946 Fr. Armanet wrote from New York to Fr. Dufault, setting off a series of events that would once again pit the vicarial superior and the Franco-Americans at the college against the long-time superior in Worcester, Fr. Martel. Fr. Armanet wanted Fr. Dufault to help him "prepare the ground" just in case a move to the Boston area might turn out to be possible. In April, as he left for France and the General Chapter, Fr. Armanet gave Fr. Dufault more specific instructions: He and one of the younger Assumptionists, Fr. Joseph Pelletier, were to set up a meeting with a Fr. Sherlock, an archdiocesan priest personally acquainted with Archbishop Richard J. Cushing.[91]

Before such a consultation could take place, Fr. Sherlock, who was about to become a member of the archdiocesan administration, indiscreetly revealed the Assumptionists' interest to Archbishop Cushing. He reported that the archbishop declared flatly that he was in favor of Assumption College's moving into his archdiocese.[92] The archbishop was also said to be disposed to "accept us just as we are," a crucial consideration for men determined to preserve the college's Franco-American identity.[93] When word of this encounter reached Paris, Fr. Armanet responded promptly. He sent to Fr. Martel, superior at the college, a letter endorsed in writing by Superior General Gervais Quénard. On May 11, Fr. Martel summoned Fr. Dufault and read him the letter by which Fr. Dufault was assigned, along with Fr. Pelletier, to seek an interview with Archbishop Cushing. They were also instructed to search for an appropriate location, with the help of whatever other priests they might need. Fr. Dufault chose Fr. Henri Moquin to assist in the project, and the three Franco-Americans decided to find a promising location before turning to the archbishop for permission to relocate in his archdiocese.[94]

Fr. Odilon Dubois, A.A.

The appointment of this exploratory committee left both proponents and opponents of relocation with the impression that the decision to move had been made, and that all that remained was to find the site and to obtain Archbishop Cushing's permission. On May 19, Fr. Martel "had a word... about what was happening regarding the development of this college" with two of the French-born professors, who were "thunderstruck."[95] Both immediately expressed their alarm in letters to Fr. Jude Verstaen, the new assistant general of the Congregation, who had served in Worcester from 1922 to 1934.

Fr. Odilon Dubois, the former subprior, had been in Worcester since 1919. He wrote that the Congregation's leaders were being fooled if they had been told that the move to the Boston area was the desire of the Worcester religious. "It is the desire of certain individuals, perhaps, who have no other care but to Americanize, to Irishize, as it is said here, and who do not see that the work we have accomplished here for 42 years will be assimilated. We engaged ourselves before the Franco-Americans to conduct a college for them, following their desire to remain what their ancestors and traditions had made them....The Franco-Americans have sustained us in the belief that we would be faithful to our promises; just recently, at the superior's feast day, the speaker reminded us of this, declaring that from the moment we were to become unfaithful to our engagements, they would simply let us fall.... May it please God that [my letter] be persuasive enough to halt the realization of projects that would be the ruin of our work."[96]

Fr. Engelbert Devinq, who had arrived in 1922, wrote that some people might be dreaming of establishing the college in another diocese, but he believed such a move would be possible only if "our spirit should change. From that moment on, our *raison d'être*, the originality that is our *raison d'être*, the friendships our faithfulness has gained for us, all will disappear.... My desire is to finish my life in the place where I have passed the greatest part of it; what would I do elsewhere? May God give me the virtue of humility and self-effacement! May he spare us the fall!!" Fr. Martel himself, who had been in Worcester since 1920, declared, "I had dreamed of another future for this house; in the fund drive...I laid out things other than those that are going to be done. I did not ask for $300,000 for that. Let someone come [from Europe], ask questions, see, inform himself and judge! At the very least, may I not be witness to this shame."[97]

Fr. Engelbert Devinq, A.A.

The anguished protests were heard and heeded in Paris. Fr. Verstaen replied immediately to Fr. Martel, with whom he had spent twelve years in Greendale, assuring him that no final decision had been made. "There is one thing on which we are in agreement a priori," he wrote to his old colleague, and "that is that the work of the Franco-American College should be continued in the sense in which it has been thus far. There are the promises made from the beginning; there are the obligations contracted with the benefactors of the college; above all there is without any doubt the fact that we had no reason to go and expend the best of our spirit and our strength in a foreign land, if we were to create there a work of a banal character that others could have created just as well. Our institution responded to a need of the Franco-American population; it should continue to respond to it, and it can do that only by preserving its originality." [98]

Fr. Verstaen also wrote that day to Fr. Armand Desautels, whose May 2 letter to Paris had stressed the urgency of authorizing someone to make an official request to the archbishop. The assistant superior general sent a set of questions, the first of which was, "Will it be possible to conduct, in Boston, a College in the Franco-American sense?" Fr. Desautels replied that the archbishop had recently had the Assumptionists notified "that the college would be admitted to the [arch]diocese ON THE CONDITION THAT IT PRESERVE ITS FRANCO-AMERICAN CHARACTER, in order to forestall the objection, in the eyes of the Irish clergy, of a competition with Boston College." Two days later, Fr. Dufault, surprised at the "doubts springing up again," wrote to Fr. Verstaen to reinforce Fr. Desautels' case. He said he could also express the support of "the great majority of the fathers (except Fr. Sup[erior], Fr. Odilon, and perhaps Fr. Engelbert." On the matter of "renouncing the Franco-American character of the college," he declared, "[I]t seems to me that we are as attached to French culture as anyone." He suggested Fr. Martel's fears on that score might arise from his actually preferring a French school over a truly bilingual one. Fr. Martel's conduct in recent years, he argued, suggested that he saw the work of the college "as destined to promote a small separatist island, making no concessions to English except inevitable ones."[99]

The business of finding a site and getting the archbishop's approval, Fr. Dufault reported, "has moved faster than we have on several occasions already, and we have been inclined to see the finger of Providence in that."[100] After a few more weeks of hesitation, on July 20 the superior general authorized Fr. Dufault and his colleagues to purchase as the new home of Assumption College a building and two-hundred-fifty acres they had found in Lexington. On July 28, Frs. Dufault and Pelletier spoke on the telephone to Fr. Sherlock about Archbishop Cushing, whom they were to meet on the following day, one day before they were to pay $10,000 for an option to buy the Lexington property. Fr. Sherlock reported that Archbishop Cushing had just returned from a trip to Canada, where he had met with one or two Assumptionists at the novitiate in Quebec. His ideas, Fr. Sherlock reported ominously, had "crystallized." An ethnic college such as Assumption could not in the long run support itself. What was needed was an American Catholic college, open to all groups.[101]

The next day, Frs. Dufault and Pelletier had lunch with Fr. Sherlock, who for some reason remained optimistic about the outcome of their afternoon meeting with the archbishop. "At two o'clock," Fr. Pelletier recorded, "the archbishop spoke to us in a very simple but very forceful way. He has two objections. He believes that the college could not support itself as it is. He does not like its 'nationalist' aspect. He will present our request to his council if we ask him to, but he will make no plea in our favor. The council will raise to him the objections he has just raised to us. If the council accepts us, so will he because we are a Catholic college and he wants to promote the religion. If we thought it worth the trouble we could write a letter presenting our request and he would discuss it with his council. We could take the option [to buy the Lexington property], but it is clear that he does not think that would be of much use."[102]

Despite that less-than-encouraging message, the Assumptionists did not consider the door shut. The next day, Frs. Dufault and Pelletier went to the scheduled meeting at a bank with their $10,000 check, ready to sign a document written in a way that would protect them if their plans were to be rejected by either religious or municipal authorities. They told their attorney and another intermediary who had been of great assistance to them about the meeting with the archbishop. Both men advised them not to purchase the option under those conditions. To the owner, his lawyer, and the representative of the bank, the priests announced that they could not sign the option agreement "for the moment." When the meeting dissolved, they drove over to the Lexington property to pick up Frs. Henri Moquin and Louis-Robert Brassard, who had been left there to take measurements in the house and elsewhere on the estate.[103]

One source of Fr. Martel's worries about the future character of the college may have been a tendency among younger Assumptionists to speculate about a future that might extend beyond the ethnic limitations of the past. The Assumptionist Franco-Americans, while they felt an obligation and a desire to take up the "perilous honor" of serving the Franco-Americans, were also acutely conscious of the position they and the college might be left in if they did not reach out to a broader clientele. On the day Frs. Dufault and Pelletier admitted they could not "for the moment" pay for an option to buy the Lexington property, they had another long discussion over lunch with Fr. Sherlock, focusing on "the possibility of winning the Franco-American clergy over to the idea of a general college with a special French department (special dormitory and courses as at Worcester currently)." Fr. Louis-Robert Brassard, once relieved of his measuring chores at the Lexington property, raised the possibility that there might be two colleges, the bilingual one in Worcester and an English-language one near Boston. Fr. Pelletier, in his journal of the "Lexington Affair," wrote that the idea Fr. Brassard had raised presented "a magnificent opportunity for the Congregation. I had already thought of it myself."[104]

The interview with Archbishop Cushing had taken place on July 29, 1946. Eighteen days earlier, Fr. Pelletier placed in his journal a brief entry for July 11: "Fr. Wilfrid makes his profession of faith and takes charge as superior at 11:45 a.m."[105] After forty-two years, the Franco-American Assumptionists had officially assumed the stewardship of the college that had nurtured them. The superior, Fr. Dufault, was a son of Spencer, Massachusetts. His subprior or assistant superior was Fr. Henri Moquin of Marlboro. The treasurer was Fr. Louis-Robert Brassard of Leominster. Perhaps to encourage successful communication across the generations, a second assistant superior was named, Fr. Engelbert Devincq from St. Omer, Pas de Calais, France.

Before leaving Worcester, Fr. Martel made a final series of gestures to reinforce his message that the college and its leaders must remain faithful to their Franco-American mission. He invited an alumnus, Fr. Adrien Verette, well-known as an ardent Franco-American, to speak at the June commencement about what Fr. Dufault, with considerable annoyance, summarized as "the great worry concerning *la survivance,* given the attitude of abandoning our culture, an attitude supposedly appearing among a good number of Francos, *even among the teachers at the college.*"[106] In a farewell message in the college magazine, Fr. Martel wrote of his many years of getting to know Franco-Americans, with whom he so identified that "when in 1935 it was deemed fit to name me as superior, I had no greater ability than any other, but I was heart and soul a Franco-American." Of his successor he wrote, "You know him. He, too, will have it at heart to follow the same paths as his predecessors, to do more and better than his predecessors, and to conduct the college to finer destinies in a spirit of unfailing fidelity."[107] Fr. Martel thus left his successor with ideological recommitments as well as financial pledges to honor.

Two months later, in his greetings to the magazine's readers, the new superior and president paid tribute to his departed predecessor, called upon friends and alumni not to forget Fr. Martel as he became superior of a war-damaged orphanage in France, then turned to current business. The addition to the high school, begun by Fr. Martel, was still being completed. The building of the new college, Fr. Dufault again explained, would have required too much in the way of materials to be possible right away; meanwhile, the current construction had allowed the admission of more college students as well as more at the high-school level; "thus the funds dedicated to this building will be reimbursed in a short time and will have received the best possible investment while waiting to be used, as is well understood, for the new college. We hope for a realization as soon as possible for this new college, the common dream of the Fathers and their friends. . . . May God and Our Lady of the Assumption help us."[108]

Notes

ABBREVIATIONS
AC — Archives of Assumption College, Worcester, MA
AG — Archives of the Assumptionist General House, Rome
AP — Archives of the Assumptionist Provincial House, Brighton, MA
PA — Archives of the Assumptionist Province of Paris
SP — Archives of the Diocese of Springfield, MA

1. Crescent Armanet to Clodoald Sérieix, March 4, 1930, AP; Armanet to Gervais Quénard, Feb. 18, 1931, AG 2PL-287; Armanet to Quénard, April 16, 1931, AG 2PL-288.
2. Armanet to Sérieix, March 16, 1930; April 26, 1930; Dec. 2, 1933, AP; Armanet, Financial Report, June 30, 1934, AG, 2PN-136.
3. *Bulletin,* (Worcester, 1933–34), 14; *Prospectus,* 1936–1937 (Worcester, 1936), 14.
4. "Archives de la maison," I, 152. AP; Armanet to Sérieix, March 21, 1934, AP.
5. Armanet, Rapport semestriel, Jan. 11, 1932, AG 2PN-79; Armanet to Sérieix, Nov. 10, 1933, Dec. 2, 1933, AP; *L'Assomption* 3, 5 (May 1934): 114.
6. *L'Assomption* 3, 5 (May 1934): 115; *L'Assomption* 3, 8 (Oct. 1934): 206.
7. *L'Assomption,* New Series 6, 4 ([no month] 1929): 22–23.
8. *Prospectus* 1933–1934 (Worcester, 1934), 10.
9. *Prospectus* 1933–1934 (Worcester, 1934), 11, 14.
10. *Prospectus* 1933–1934 (Worcester, 1934), 18–24.
11. *Prospectus* 1933–1934 (Worcester, 1934), 27.
12. *Prospectus* 1933–1934 (Worcester, 1934), 31–38.
13. *Prospectus* 1933–1934 (Worcester, 1934), 38–40.
14. *Prospectus* 1933–1934 (Worcester, 1934), 38–47.
15. *Prospectus* 1933–1934 (Worcester, 1934), 28.
16. *Prospectus* 1933–1934 (Worcester, 1934), 29.
17. *Prospectus* 1933–1934 (Worcester, 1934), 29; on finishing without the degree see, for example, Armanet to Sérieix, June 5, 1934; Armanet to Bal-Fontaine May 31, 1935, AP.
18. *L'Assomption* 2, 9 (Sept. 1932): 147; Sérieix to Maubon, Jan. 25, 1925, AP.
19. Clodoald Sérieix, "College de *L'Assomption,* Worcester, Mass. Notice sur sa fondation et sa situation actuelle, fevrier 1928," typescript, AP (emphasis in original).

20. Armanet to Sérieix, Sept. 21, 1933, Sept. 10, 1934, AP; *L'Assomption* 4, 18–20 (June, July, Aug. 1937): 507. The men pictured were Frs. Clair Boisvert, Henri Moquin, Leo Martin, Wilfrid Dufault, Antonio Laberge, Pierre-Rodolphe Martel (a Canadian), Ulric Charpentier, Edouard Moreau, and Armand Desautels. Absent were Frs. Marcellin Parent and Leopold Braun.

21. Armanet to Sérieix, Jan. 13, 1934, AP; Armanet to Sérieix, Feb. 11, 1934 AP; Richard Richards, A.A., *The Assumptionists,* 90–91.

22. Henry Moquin, A.A. and Richard Richards, A.A., *Assumptionists in the United States,* 101; Armanet to Sérieix, June 4, 1932, AP.

23. Armanet to Quénard, April 26, 1934, AG 2PL-293; Buisson to Sérieix, April 13, 1934, AP.

24. Armanet to Sérieix, Feb. 11, 1934 AP (emphasis in original); Armanet to Quénard, April 26, 1934, AG 2PL-293 (emphasis in original); Report of Diocesan Consultors, Sept. 27, 1934, manila folders, SP.

25. Armanet to Bernardin Bal-Fontaine, April 25, 1935, AP.

26. Armanet to Sérieix, Feb. 11, 1934, March 2, 1934; Armanet to Bal-Fontaine, April 25, 1935, AP; Armanet to Kokel, April 25, 1935, AG 2PK-98; Armanet to Quénard, May 31, 1935, AG 2PL-297; "Archives de la maison," II, 26.

27. Rodolphe Martel to Quénard, Feb. 5, 1936, AG 2PL-300.

28. *L'Assomption* 3, 22 (Oct. 1935): 548.

29. Martel to Bal-Fontaine, Dec. 11, 1936, AP.

30. Armanet to Sérieix, July 28, 1930, AP; Armanet to Sérieix, March 28, 1931, AP; Armanet to Sérieix, June 28, 1932, AP.

31. *L'Assomption* 3, 6 (June 1934): 177.

32. Sérieix to Armanet, March 15, 1930, PA 1930/24; Armanet to Sérieix, March 28, 1931, AP.

33. Armanet to Sérieix, March 28, 1931 AP; Armanet to Sérieix, Aug. 27, 1934, AP; *L'Assomption* 3, 6 (June 1934): 175–78.

34. Armanet to Quénard, April 16, 1931, AG 2PL-288; Armanet to Sérieix, Dec. 28, 1932, AP.

35. Armanet to Sérieix, March 6, 1932, AP; see also Armanet to Sérieix, Jan. 5, 1932; Armanet to Journet, March 10, 1932, AP.

36. Armanet to Quénard, April 22, 1930, AG 2PL-285; Armanet to Sérieix, Jan. 5, 1932, AP; *Lettre à la dispersion,* no. 427 (1933), 53–54.

37. Armanet to Quénard, Feb. 22, 1933, AG 2PL-289.

38. Armanet to Sérieix, April 30, 1934, AP; Statement of Expenses, July 1935, AP.

39. Armanet to Quénard, Feb. 22, 1933, AG 2PL-289.

40. Armanet to Remi Kokel, April 25, 1935, AG 2PK 98.

41. Armanet to Quénard, May 31, 1935, AG 2PL-297.
42. *L'Assomption*, New Series 1, 5 (May 1931): 250.
43. *L'Assomption* 4, 24 (Dec. 1937): 631.
44. Armanet to Quénard, Dec. 15, 1933, AG 2PL-292.
45. *L'Assomption* 3, 6 (June 1934): 161–62.
46. Armanet to Quénard, May 31, 1935, AG 2PL 297.
47. Martel to Bal-Fontaine, June 13, 1935, AP.
48. Martel to Bal-Fontaine, May 26, 1937, AP.
49. Martel to Quénard, Feb. 5, 1936, AG 2PL-300; Martel to Quénard, March 30, 1936, AG 2PL-301; Martel to Bal-Fontaine, May 26, 1936, AP.
50. Martel to Bal-Fontaine, June 9, 1936, AP.
51. Martel to Bal-Fontaine, July 6, 1936, AP. The estimate in dollars is based on Fr. Martel's January 1936 financial report, in which he used an exchange rate of 15 francs to the dollar (AG, 2PN-89).
52. Martel to Bal-Fontaine, Oct. 21, 1936, AP.
53. Martel to Quénard, Feb. 5, 1936, AG 2PN-89; Martel to Bal-Fontaine, June 9, 1936, AP; *L'Assomption* 4, 9 (Sept. 1936): 215, 193.
54. Martel to Quénard, n.m., n.d., 1937, AG 2PK-112.
55. Martel to Bal-Fontaine, June 9, 1936, AP.
56. "Archives de la maison," I, 152, AP.
57. "Archives de la maison," I, 151, AP; Guissard, *Un demi-siècle d'histoire, 1904–1954* (Worcester, 1954), 157; Antonio Laberge to Felicien Vandenkoornhuyse, April 15, 1941, AG 2PK-117.
58. "Archives de la maison,"I, 151, AP.
59. "Archives de la maison," II, 191, AP, and Guissard, *Un demi-siècle d'histoire*, 1904–1954, 157, state that the entire debt had been retired as of January 1945. Martel to Bal-Fontaine, April 8, 1946, AP, shows that the $70,000 was still outstanding over a year later.
60. House Council Minutes, p. 16, AP. The amount forgiven may have been $5,000 rather than $15,000. It is not clear from the manuscript whether the clerk intended to write a "5" partially obscuring a previous vertical stroke or whether the stroke is a "1" intended to precede the "5." Fifteen thousand seems more likely.
61. "Archives de la maison," II, 138, 152, 159, 186, AP; Armanet to Quénard, Nov. 18, 1944, AP.
62. *L'Assomption* 6, 13 (Jan. 1941): 305-11.
63. *L'Assomption* 8, 21–22 (Sept.–Oct. 1945): 348–50.

64. *L'Assomption* 21–22 (Sept.–Oct. 1945): 348–50; Armanet to Quénard, Nov. 18, 1944, AP.

65. *L'Assomption* 9, 5 (Oct. 1946): 98.

66. Laberge to Vandenkoornhuyse, April 15, 1941, AG 2PK-117.

67. *L'Assomption* 4, 21 (Sept. 1936): 199.

68. "Archives de la maison" II, 125, AP, *L'Assomption* 6, 18–20 (June, July, Aug. 1941): 450.

69. *L'Assomption* 6, 18–20 (June, July, Aug. 1941): 450.

70. *L'Assomption* 6, 5 (May 1940): 106.

71. *L'Assomption* 7, 17–18 (June 1943): 328.

72. *L'Assomption* 7, 19–20 (Aug. 1943): 357 (emphasis in original).

73. *L'Assomption* 8, 15–16 (March–April 1945): 250–51.

74. *L'Assomption* 8, 11–12 (Nov.–Dec. 1944): 164, 185–86.

75. *L'Assomption* 8, 11–12 (Nov.–Dec. 1944): 164, 185–86; "Pour le College de *L'Assomption*," brochure, AG 2PK-118.

76. Martel to Jude Verstaen, June 13, 1946, AG 2PK-123.

77. "Archives de la maison," II, 157–65, AP; Dufault to Verstaen, June 11, 1946, AG 2PK-122; Dufault, "Notes on the Relocation of Assumption College: 1940s and 1950s," typescript, 1991, AC.

78. Martel to Verstaen, June 13, 1946, AG 2PK-123.

79. Between March 27, 1944, and November 1945, the minutes record only one meeting, on October 18, 1944, and building plans were not discussed on that occasion. "Council Minutes, 670 West Boylston [Street]. Worcester, Mass., 1932–1956," 15–16, AP.

80. Dufault to Verstaen, June 11, 1946, AG 2PK-122.

81. Martel to Verstaen, June 13, 1946, AG 2PK-123; Dufault to Verstaen, June 11, 1946, AG 2PK-122; *L'Assomption* 8, 13–14 (Jan.–Feb. 1945): 193.

82. Armanet to Quénard, Nov. 18, 1944, AP.

83. *L'Assomption* 8, 13–14 (Feb. 1945): 197; Gerard Brassard to Bal-Fontaine, May 25, 1945, AP; "Archives de la maison," III, 25, AP; *L'Assomption* 8, 17–18 (June 1945): 258–59.

84. Martel to Verstaen, June 13, 1946, AG 2PK-123.

85. Armanet to Bal-Fontaine, April 2, 1945, AP.

86. "Archives de la maison," III, 16–17, AP; Armanet to Bal-Fontaine, Sept. 20, 1945, AP; *L'Assomption* 8, 21–22 (Sept.–Oct. 1945): 337.

87. *L'Assomption* 8, 17–18 (May–June 1945): 259; 8, 21–22 (Sept.– Oct. 1945): 371, 322; 8, 23–24 (Nov.–Dec. 1945): 354.

88. *L'Assomption* 8, 21–22 (Sept.– Oct. 1945): 337; Martel to Quénard, Dec. 7, 1945, AG 2PL-309; Martel to Quénard, Jan. 1, 1946, AG 2PL-313.

89. Dufault to Verstaen, June 11, 1946, AG 2PK-122. In this letter, Fr. Dufault referred to the October 1945 report to Fr. Quénard.

90. Martel to Quénard, Dec. 7, 1945, AG 2PL-309; Dufault to Verstaen, June 11, 1946, AG 2PK-122; Martel to Verstaen, June 13, 1946, AG 2PK-123.

91. Dufault to Verstaen, June 11, 1946, AG 2PK-122.

92. Dufault to Verstaen, June 11, 1946, AG 2PK-122.

93. Armand Desautels to Verstaen, May 2, 1946, AG 2PK-119.

94. Dufault to Verstaen, June 11, 1946, AG 2PK-122; "Archives de la maison," III, 39, AP; Verstaen to Martel, May 28, 1946, AG 2PK-121.

95. Martel to Verstaen, May 20, 1946, AG 2PK-119.

96. Dubois to Verstaen, May 19, 1946, AG 2PK-119.

97. Devincq to Verstaen, [May 19?], 1946, AG 2PK-119; Martel to Verstaen, May 20, 1946, AG 2PK-119.

98. Verstaen to Martel, May 28, 1946, AG 2PK-121.

99. Verstaen to Desautels, May 28, 1946, AG 2PK-120; Desautels to Verstaen, June 9, 1946, AG 2PK-122 (emphasis in orginal); Dufault to Verstaen, June 11, 1946, AG 2PK-122.

100. Dufault to Verstaen, June 11, 1946, AG 2PK-122.

101. Dufault to Verstaen, June 11, 1946, AG 2PK-122; Joseph Pelletier, "Efforts to Move College to Boston Area," typescript, AP, 5–7.

102. Joseph Pelletier, "Efforts to Move College to Boston Area," typescript, AP, 5–7.

103. Joseph Pelletier, "Efforts to Move College to Boston Area," typescript, AP, 5–7.

104. Pelletier, "Efforts to Move College to Boston Area," typescript, AP, 7.

105. Pelletier, "Efforts to Move College to Boston Area," typescript, AP, 5.

106. Dufault to Verstaen, June 11, 1946, AG 2PK-122 (emphasis in original).

107. *L'Assomption* 9, 4 (Aug. 1946): 73–74.

108. *L'Assomption* 9, 5 (Oct. 1946): 97.

Chapter Four
Winds of Change: The Tornado and the "New College" (1946–1964)

As they accepted responsibility for Assumption College and its future, the Franco-American Assumptionists inherited both the plan to put up college buildings on Burncoat Street and the worrisome mission of providing a French-based Catholic education to a diminishing population of young French-speaking males. As we shall see, it was the problem of maintaining the institution's French identity that soon preoccupied the superiors and their personnel. While they were grappling with the cultural and curricular implications of opening their doors to a wider clientele, a tornado battered the college on June 9, 1953. The great storm put an end to the Burncoat Street plan, and it inspired a fresh search for a location that Assumption College — separated from the high school (renamed Assumption Preparatory School) — could call home. Already in the midst of a difficult change before the tornado, the Assumptionists and their lay co-workers set out on an uncharted sea that would carry them, and their evolving institution, to Salisbury Street.

When the Franco-American era at Assumption began, Fr. Wilfrid Dufault was leading a religious community of fifty-one men, among whom were twenty-eight priests, fourteen lay brothers, and nine choir brothers (aspirants to the priesthood). Of the hundreds of Europeans who had served the college over its first four decades, only seven (five priests and two lay brothers) remained. In September 1946, as so often in the past, the start of the school year had to be postponed (twelve days for the high school, nine days for the college), due to the construction still under way over the gymnasium. When students finally arrived, the religious and their lay associates welcomed 406 students to the high school and 93 to the college. Among the college students, there were only 5 in the junior class and 2 in the senior class.[1]

The first Mass in the still unfinished Chapel of Christ the King was celebrated on Nov. 9. Preparations for the eventual construction of a "new college" on Burncoat Street also proceeded. The garden of the college's farm was transformed into playing fields that were to separate the two schools, and the superior with his house council decided to rid themselves of the farm operation as soon as possible. Lingering hope for an alternative location for the college led the house council in November 1946 to approve one more letter to Archbishop Cushing seeking permission to locate in his archdiocese. Despite that gesture, when Fr. Dufault filed his first report to Rome in January 1947, he reported the purchase of additional land on Burncoat Street.[2]

Fr. Dufault's presidency, which had begun in August 1946, came to an early but not unexpected end in January 1947. On January 2, the vicar provincial, Fr. Crescent Armanet, sent word that the vicariate of North America, which had depended on the Province of Paris since 1923, had been elevated to the status of an autonomous province, and that the first superior of the new province would be Fr. Dufault. In March, Fr. Henry Moquin, dean of studies at the college since 1935, vice president and subprior since the previous August, was appointed the new superior in Worcester. As provincial superior. Fr. Dufault continued to be based at the college until September 1950, when he moved his office to Our Lady of Guadalupe parish in New York City.[3]

Named as the new subprior was the energetic and resourceful Fr. Armand Desautels, who had been serving as dean of studies. Along with Frs. Dufault and Moquin, Fr. Desautels had long been regarded as an important leader among the first generation of Franco-American Assumptionists. The three men were entering a new and important official partnership that would help to sustain the college as it confronted staggering challenges in the decades ahead. A key to their successful collaboration may have been the attitude expressed by Fr. Desautels to his superior general. "I will do my best," he wrote, "to deserve the confidence you are placing in me — understanding very well, however, what I lack for carrying out this office.

"What encourages me," he added, "is my confidence in my Superiors. I will be working with Frs. Wilfrid and Henri; it will naturally be easy to put all my heart into that. It has been a long time since I learned to value them — it goes back to the days when we were students together, at the College. The years since then have only strengthened my feelings for them. So I can say in all sincerity that it feels good to work in concert with them for the Assumption [the Assumptionist mission]."[4]

At the time of his appointment as superior, Fr. Henri Moquin was deeply involved in negotiations that would eventually bring to the campus two military surplus buildings (permanently labeled "barracks" by students and faculty alike) to serve as classrooms and laboratories for the college-level students. These acquisitions were to supplement collegiate housing secured by the purchase of four "three-deckers" adjacent to the college on Assumption Avenue and Randall Street. When he was feted in May by the religious community, Fr. Moquin spoke of his worries about the college's finances, which were unable to keep pace with the rising cost of living, despite several increases in tuition and fees. The school was running a deficit, he confided, and the problem was due most of all to the large number of lay professors, whose salaries were swallowing up any profits that might have been realized. Tuition would have to go up again, and there was no sense hoping for more religious personnel, given the new foundations the young province was envisioning. He urged his religious to seek divine assistance through prayer and through the faithful observance of their religious life.[5]

Lay professors were not the only expensive employees. Fr. Moquin also worried about the cost of workers hired to do some of the physical labor on the campus. In February 1947, the house council decided to turn to oil rather than coal as a heating source so that lay brothers could be released from tending fires and assigned to replace hired workers. In his first report to his superiors, where the standard form asked what was his most urgent need, Fr. Moquin answered, "religious professors and lay brothers." Year after year, he answered the question the same way, sometimes leaving out the lay brothers and confining himself to a plea for "religious with degrees."[6]

The challenges facing Fr. Moquin and his colleagues were complex and interrelated. The number of college students would have to increase if the institution was to regain its equilibrium and prosper, but adding students meant adding lay faculty and creating what Fr. Moquin called "a big hole in the budget." To make ends meet, he tried raising tuition and fees, but that led to lower enrollments, especially in the high school, which was the principal source of future students for the college. The high school population shrank from 396 students in 1947 to 188 in 1951. Part of the budgetary problem was caused by the need to improve library holdings and laboratory facilities in order to gain accreditation. That goal was achieved in December 1949, when the college-level program was approved by the New England Association of Colleges and Secondary Schools. However, the college population, which had increased from 93 students in 1947 to 179 in 1949, declined to 154 in 1951 and to 140 in 1953.[7]

How French? How Classical?

The uphill struggle to attract more students encouraged many in the administration and the faculty to consider changes that had once seemed beyond discussion. In his 1949 annual report to

Rome, Fr. Moquin wrote: "It is to be noted that the professors are not in agreement on the necessity of preserving the classical curriculum, nor on the necessity of insisting on French." In 1951, he gave two important cultural reasons for the declining number of students: "a diminution of interest among the young Franco-Americans for French, and especially a general disinterest in a classical curriculum. Courses that prepare directly for careers are preferred." He noted that the Assumptionists' "neighbors, the Jesuit, Dominican, Augustinian, Holy Cross and Franciscan fathers have introduced courses [of study] parallel to the classical. Barely a third of their students are following the classical. In imitating them we would not have fewer students in the classical, and we would reach a larger number of young Catholics."[8] Among Assumptionists in Worcester, the image of the college's clientele was expanding to embrace "young Catholics" in general, rather than just Franco-Americans. At the same time, the commitment to a classical curriculum, with its emphasis on Latin, Greek, French, religion, history, literature, and philosophy, was being challenged by demands for a more "practical," career-oriented curriculum.

In 1952, the religious community in Worcester met to deal with proposals that reflected their sense that changing times required changes in the high school and the college. Although the institution had on occasion admitted students who were not Franco-American by birth, in 1952 the community decided that, for the first time, the high school and college would actively recruit students who were neither French speakers nor ethnically Franco-American. In one of his last tasks as superior and president, Fr. Moquin described that change, and the thinking behind it, in the college's magazine, *L'Assomption.* In a tone more rueful than celebratory, Fr. Moquin observed that there had been "great voids" appearing in the college's dormitories and study halls because Franco-American young men were demonstrating increasing "indifference" toward French culture. "In the face of these facts and in the hope of saving a center of French culture, at least for the small number who are still sufficiently interested in it to send us their sons, it seemed to us necessary to introduce side-by-side with our current courses [of study], in which the vehicular language is French, other courses in which the vehicular language will be English. To these courses we will admit all who present themselves, provided they wish to follow the program. We also nurse the hope of being able to enlist among the students non-Franco-American young men who wish to acquire the elements of French culture."

"We take this action under economic pressure and out of a desire to survive."

In the high school, Fr. Moquin added, one of the new courses of study would be designed for students interested in pursuing studies in business administration when they went on to college. This decision, too, was based on "the needs of our population which, like the others, is withdrawing more and more from the classical [curriculum]."[9]

The college's movement away from ethnic exclusivity and from the classical curriculum had barely begun when word arrived in May 1952 that the provincial superior, Fr. Wilfrid Dufault, had been elected superior general of the Assumptionist Congregation around the world. As Fr. Dufault moved from New York to Rome, his place as provincial superior was taken by Fr. Henri Moquin, and Fr. Armand Desautels was appointed superior in Worcester and president of the college. As local superior, provincial superior, and superior general, the trio of Franco-American friends would soon find themselves making momentous decisions about the fate of the college under circumstances none could foresee.[10]

Fr. Desautels had provided strong leadership as dean of studies and had been an important contributor to an emerging consensus that the college had to "broaden its framework" both ethnically and academically. This spirit of openness was reflected in a new practice that began in 1952, when the house council approved a proposal from Fr. Desautels to set up a "lay council," made up of businessmen, to provide continuous advice to the

Fr. Armand Desautels, A.A.

superior regarding financial matters. Archibald Lemieux, president of Wright Machine Company in Worcester, was designated as the first member. Already a trusted advisor, Mr. Lemieux had in 1946 established a scholarship fund at the college in memory of his parents. After recording in the minutes the establishment of the lay council, Fr. Desautels added: "*Budgets.* Before allowing spending on all sorts of things, begin by establishing a budget. Do that with the idea that there is construction to be done that will be expensive. I will do nothing in this area without a council of people who know all about business." Although the practice of seeking lay advice stretched back to the beginning of the college, from 1952 on the college would have a formally constituted body of outside advisors, mostly businessmen, who were regularly consulted on financial matters, and sometimes on matters of educational policy as well.[11]

As he began his first academic year as superior and president, Fr. Desautels called a special meeting of the house council, along with a selection of other Assumptionists, to begin addressing "the role of French" at Assumption now that the community had faced the practical reality that there were too few Franco-American students "to allow the college to develop normally." On "the question of principle," Fr. Desautels reported: "We believe that Assumption, in the future as in the past, should base its individuality and its reputation on French. Among Catholic colleges, that is what distinguishes it. We have the formation of serious Catholics as our first purpose, but that is common to all Catholic colleges. We have a good college education to offer, but that is common to all colleges. What distinguishes us is French."

The participants also agreed that "an intense French culture" should be offered to those who came looking for it, but that did not mean, Fr. Desautels noted, that all students should be expected to pursue French intensively or that French had to continue to be the "vehicular language" of the institution. The meeting found itself searching for ways in which the French cultural heritage, especially the Catholic dimension, might be conveyed in English. The college might, for example, publish "a review whose purpose is to give in English the values of French culture, and particularly its Catholic values." Fr. Desautels may have been papering over disagreements about the future place of French in the curriculum when he noted agreement on the proposition that, since language is "inseparable from culture, it is necessary for us to make it *well understood*."[12]

The superior consulted others, including the man who in 1950 had become the first bishop of the newly created Diocese of Worcester. Bishop John J. Wright was an ardent admirer of French Catholic culture and an important source of advice and support for the Assumptionists and their college. He evidently had considerable influence on Fr. Desautels' next effort to help his community redefine the college and choose the path it would follow as it moved away from its historic role as the academic and spiritual incubator of a Franco-American elite. To help move the discussion along, Fr. Desautels circulated among all the priests at the college a summary of the "very frank and very free discussions" that were under way.

Among "essential principles" upon which all were agreed, he listed the college's Catholic identity, its commitment to first-rate education, and the fact that it had its own "personality." That personality had three dimensions: It was Augustinian, Assumptionist, and "oriented toward the classical [curriculum], toward languages, and in a very particular manner, toward French." He did not define Augustinian, but did write that the college, due to its Assumptionist character, "should lead the world toward the love of Jesus Christ, of the Blessed Virgin, of the Church; it should develop liturgical worship, [and carry on] the fight against the 'revolution' in the sense of Fr. d'Alzon."

If the college's goal was merely to become a large Catholic college, Fr. Desautels wrote, then, "Bishop Wright tells us, we will follow the path of many colleges that have lost their former personalities."

Fr. Desautels closed with "practical conclusions" on two points: the college's French character and its future location. "It is suggested to us," he wrote, without identifying a source, "that the French aspect of Assumption be distinguished by a name, such as the Institute of French Culture, which might become nationally associated with Assumption College. The Institute would provide a full curriculum in French, all the way to a doctoral degree, which, it was hoped, might become well known. This curriculum would give a major role to French Catholic thought."

In 1950, the college's state charter had been revised to allow the conferral of graduate degrees. The immediate objective had been to provide teachers of French the opportunity to improve their skills while taking courses leading to the master's degree. The program began with noncredit evening courses, followed by a summer session in 1952 that marked the beginning of graduate education at the college. The early success of that endeavor encouraged new aspirations. The suggested Institute of French Culture would seek to promote "an intense French social life" comparable to the one already enjoyed during the summer sessions. This "French life should penetrate the entire college at certain periods of each day," something that would become more and more accepted as students of varying backgrounds were attracted to the college precisely for its French cultural life.

New departments, Fr. Desautels continued, should be added only when they could accommodate themselves to the distinctive personality of the college. "For example, Bishop Wright says to us, a curriculum in 'Business Administration' is imperative, but not in the way conceived in neighboring colleges; only what is common to all and judged essential would be borrowed, in order to orient the other courses toward the legal aspect of business and relations with France and other countries. For example, he [Bishop Wright] continues, a 'pre-diplomacy' department would fit entirely with our purpose."

Where should the "new college" with its French Cultural Institute be located? Despite being "close to Holy Cross," it would make sense for Assumption to remain in Worcester, provided it was distinctive, and not just another Catholic college. Financially speaking, it was impossible to raise the $2 million or more that would be needed to construct a campus elsewhere in Worcester, so the only realistic course was to carry out the plan to build on the Burncoat Street side of the property. This meant, Fr. Desautels concluded, that Assumption would never become a large college. He estimated that a Burncoat Street campus could accommodate up to 700 college students.[13]

For the immediate future, Fr. Desautels and the faculty concentrated on revising the curriculum. Under the existing requirements, all students took many courses in common, but options, mainly in the final two years, allowed them to pursue courses of study labeled prelegal, preeducational, premedical, and, beginning in 1948, pretechnical. In early 1953, the president carried his effort to describe and explain the changing college into the institution's magazine. To his own question, "What is Assumption?" he provided a three-part answer: a Catholic college, a liberal arts college, and a college that spreads French culture.[14] He then announced that the faculty had developed a new educational plan in keeping with this definition. Doing away with the old career-linked titles, the new system would place heavy emphasis on general education. Unless they had to take courses prerequisite to their future "concentrations," students during their first two years would take only general education courses in English, literature, Western civilization, physics or chemistry, French, and a second foreign language chosen from Latin, Greek, Spanish, German, and Russian.

During the junior year, students would continue to devote half their course work to general education courses in theology, philosophy, and French literature (in English, for students lacking adequate skills in French), supplemented by an interdisciplinary reading course intended to allow concentrators in a given field to meet in small groups "to build up and coordinate background and to begin evaluating it critically." In the final year, the general education requirements in theology and philosophy were to be supplemented by an interdisciplinary coordinating seminar designed to help students connect their learning from various courses taken during their four years. The program would conclude with a final comprehensive examination to "test the student's assimilation of and his ability to use all of his acquisitions." Students could concentrate in English, French, philosophy, natural sciences or a new field, introduced into the college for the first time, business administration.[15] The new curriculum, which came to be called the concentration-integration system, was to take effect in the fall semester of 1953.

"Like the End of the World"

After it had been discovered that the 1952–53 school year was scheduled to last a week longer than legally required, the calendar was altered to save money, and students went home a week earlier than originally expected. That turned out to be, in the words of Fr. Henri Moquin, "a curious providential event."[16] At 5:08 p.m on the afternoon of June 9, two days after graduation, while the religious were on retreat, Fr. Polyeucte Guissard looked out from his third-floor room and saw signs of a gathering storm. He closed his windows and went through the nearby dormitory and other rooms doing the same thing. When he reached the first floor, he headed for the large chapel, where a session of the retreat was scheduled for 5:25. "When I got to the door," he reported, "I was literally driven back against the wall and [I] searched for something to hold onto. I managed to open the door of an indoor telephone booth and pulled Bro. Arthur [Blais] in. I stayed there in the midst of an uproar like the end of the world. . . .

Fr. Polyeucte Guissard, A.A.

Fr. Engelbert Devinq, A.A.

When we went out, we all were running everywhere, bewildered: almost everyone with bloody faces, many without knowing it.

"An indescribable scene: [There was] no square, no tree, no garage, no farm, no college classrooms [the old wooden 'barracks']. . . . Grotto intact, the convent of the sisters vanished. . . . We rushed toward the sisters. Most of them were preparing our dinner [in the main brick building], which is what saved them. Five, I believe, were in the convent. The youngest was dug out dead after an hour or two, using whatever makeshift means we could, because no one [from outside] could reach the college; the roads were blocked. The tower fell into the boilers, which began to burn, and no water! There was a head count. All of a sudden it was realized that no one had seen Fr. Engelbert [Devincq]. People went to his room and saw that it had fallen to the lower floor, at the bottom of the entrance parlor, in an indescribable tangle of beams, etc. A search for him was begun, but before long we were obliged to leave, because it was feared that the explosion of the boilers might blow us all up. When it became possible, the search continued. He was given extreme unction in full consciousness, because he squeezed a hand. But he did not speak."[17]

Except for the hand with which he squeezed, Fr. Engelbert Devincq was still buried when he was anointed by Fr. Louis Dion. Fr. Louis-Robert Brassard was also alive in the rubble, his leg broken in two places, his face cut badly. According to the account inscribed in the "Archives de la maison," not everyone obeyed the order to leave because the place might blow up. [The two priests] were painstakingly recovered from the ruins without interruption, despite the order to evacuate the building, threatened with an explosion of the furnace and of the oil tank close to the fire. Frs. Louis Dion, Clair [Boisvert], and [Vincent] Dolbec continued their labor to extricate the two victims after having mutually given one another absolution. Fr. Dion then absolved and gave extreme unction to Fr. Engelbert, who was still conscious enough to squeeze his hand. But his heart soon stopped and he rendered his soul to God. Death claimed two other victims: Sister Marie St. Jean Martel, 26 years old, and Sister Marie St. Hélène Simard, 48 years old." Both were members of the Antonian Sisters of Mary.[18]

Antonian Sisters of Mary waiting for help

When the tornado struck the college in Greendale, the three principal leaders of the Franco-American Assumptionists were away. Fr. Armand Desautels, along with the provincial superior, Fr. Henri Moquin, had left that morning for Quebec, while the superior general, Fr. Wilfrid Dufault, was in France. When Frs. Desautels and Moquin heard the news on the radio at 11 p.m., they turned around and headed back to Worcester. Within days, Fr. Dufault had arrived as well. In the midst of an outpouring of public sympathy and financial support, Fr. Moquin and his colleagues, at a meeting of the provincial council, decided that the Greendale building would be restored to serve as a high school (renamed Assumption Preparatory School) and that the college would be transferred to some other site. As the summer advanced, most of the religious community, along with many other volunteers, helped take apart the mountain of rubble and clean up and restore the buildings. Fr. Moquin transferred his residence from New York to Worcester and worked closely with his architects, Normand and Albert Nault, to design the restoration. The president, Fr. Desautels, traveled throughout New England and into Canada seeking contributions to the Assumption College Reconstruction Fund. By October, Assumption Preparatory School, though still under

Before

reconstruction, could open its doors to 262 students.[19]

Among the many who contributed to the survival and recovery of the college, two local academic institutions played especially critical roles. On the night of the tragedy, the Jesuits at the College of the Holy Cross opened their residence to the Assumptionists. They "took in our homeless community," Fr. Dufault later wrote, "and gave food and shelter to 50 or 60 religious for the entire summer of '53 while the Greendale campus was being rebuilt." Holy Cross also provided a classroom building for the use of the Assumption summer program. When Assumption went into temporary quarters at 1010 Main Street in the fall of 1953, nearby Clark University allowed the students and faculty the use of its library as well as some laboratory and athletic facilities.[20]

After

The main college building (now the prep school building) had been insured, but not against wind damage. That unpleasant fact was somewhat offset by the fact that the college — the eight-year institution — at the time of the disaster was in sound fiscal condition. Despite Fr. Henri Moquin's worries about a deficit in 1947, the college had been operating in the black since 1950 and had managed to free itself of debt by mid-1951.[21] Fr. Moquin had also obtained permission, in June 1952, to borrow $230,000 to help build a $450,000, 105-room dormitory for college students, three-fourths of whom were still living in wooden structures purchased in 1946. The new building was intended principally to attract more students, but Fr. Moquin noted that putting it up would also put an end to the criticisms that had emerged after the last fund drive, the 1945–46 campaign, when Fr. Rodolphe Martel had devoted the proceeds to the new chapel, dormitories, and other rooms built

1010 Main St.

above the gymnasium, rather than to the construction of a promised "new college."[22] Fortunately, when the tornado struck a year later, construction of the new college residence on the projected Burncoat Street campus had not begun.

While searching for a permanent home, the college in July 1953 purchased as its temporary campus the land and building at 1010 Main Street that had been the home of the Worcester School of Business Science. There would be 101 lay students taking courses at the college on Main Street, plus 16 Assumptionist scholastics (students). Lodging for the lay students was arranged with families in the surrounding neighborhood, while the scholastics commuted with the religious professors from and to the campus in Greendale. In September, Fr. Desautels reported to Fr. Dufault that an eighty-acre site for a permanent campus had been found on Salisbury Street, not far from the Ellis Estate they had built dreams on in 1943. It was, he reminded his colleague, "in a magnificent section of the city," and he noted that an adjacent

Aerial view of new college site

site on Nelson Place might also be available. By March 1954, the college magazine was able to report the purchase of a "95-acre tract of land" including "69 acres on the northeast side of Salisbury Street, and 26 acres off Nelson Place."[23]

Another location that had been seriously considered was in northeastern Worcester, at the city of Worcester's former Home Farm for destitute people. The main problem there, Fr. Desautels said, was time. It did not seem that it would be possible to open the college there in 1955, a "very serious" consideration. Fr. Desautels' eagerness to get a new campus started was based largely on the desire of the college students for a residential milieu of their own, but also on the college's need to retain and attract students. "In the course of this year," the superior and president remarked in July 1954, "there has been practically no 'college life,' the students have learned to love what they don't have, and [they] are so eager to return to places of their own on a new campus that a large number of them would not consent to stay with us if they were not assured that their house would soon be built." He had also made something of a moral commitment to the preparatory school seniors living and studying in Greendale. "Many of them," he wrote, "who were making plans to go somewhere other than to us for college, changed their minds the day they were shown the [architect's] model of our dreams for the future, and were told that we hoped to see a first building ready for September 1955."[24] The hope turned out to be premature.

The college's temporary home, the Main Street land and building, had been purchased for $70,000, and the vacant Salisbury Street property had cost $69,000. Most of the reconstruction fund of around $1 million had been spent on the rebuilding of what had become Assumption Preparatory School. When the new convent for the Antonian Sisters was completed in Greendale, there would be around $100,000 left in hand — and a college yet to build. As was his custom, Fr. Desautels consulted widely among businessmen, including consultants who gave advice on fund drives.[25]

Fr. Armand Desautels, A.A. and Bishop John J. Wright with architect's model of new cam-

Bishop John Wright, who had become chairman of the president's Advisory Council, made one of his frequent visits to the Greendale campus to preside over the June 1954 commencement ceremonies. On that day, Fr. Desautels told him that he had received a caution about conducting a major drive in the near future, when the reconstruction campaign had barely finished. The consultants suggested that a drive should be preceded by a year of public relations intended to make the college and its mission better known. Bishop Wright wrote a few days later to say that he had "thought this over, carefully and prayerfully," and had "reached the conclusion that there is a great wisdom in the recommendation. . . . I am very fearful that the prestige and perhaps even the future of the College might be jeopardized by any fund raising appeal which did not click and which left us in the position of seeming to be discredited for future like appeals." The bishop asked Fr. Desautels to schedule a meeting of the advisory board in August on a day when "I can certainly be present" to discuss the matter.

Bishop Wright closed with personal advice that testified to the bond between the two men. He urged the president to "get plenty of rest" in July. "You have borne up magnificently under the stress of your work," he wrote, "but there are limits to what you can endure. I beg you not to get too close to those limits.

"The College is retaining its continuity and its existence in the temporary headquarters. Haste makes waste — and in this case, could spell death." [26]

Reaffirming French Identity

Even if the tornado had not occurred in 1953, any history of the college would have to treat the early 1950s as a turning point in the institution's life. The decision in 1952 to open the high school and the college to students who were not Franco-Americans, and the announcement in 1953 that a new "liberal arts" curriculum, with an optional concentration in business administration, would take the place of the old classical course of study were major departures from the college's traditional ways of thinking about itself and pursuing its mission. These changes elicited passionate objections from members of the Franco-American press and the Franco-American societies that had always been among the college's most ardent supporters. News that the college was no longer dedicated exclusively to the education of a French-speaking Catholic elite elicited cries of betrayal from people who had devoted a great deal of labor and money to sustaining such an institution. Few could resist reminding the Assumptionists of the college motto, *Je maintiendrai* [I will uphold].

One of the most acerbic critics was a Worcester writer who had long suspected that the Franco-American Assumptionists would be disloyal to their French heritage. Wilfrid Beaulieu, editor and publisher of *Le Travailleur,* muted his comments for a few months after the tornado and published materials supporting the reconstruction effort. On October 1, 1953, however, his restraint ended. The favorable publicity the college was receiving, he wrote, "unfortunately attaches itself like a bloodsucker to symptoms of anglicization that are more than worrisome when it comes to the truly bilingual future of this college that for so long was called 'the college of the Franco-Americans.' By means of the tornado and of the disaster suffered by the college, it very much seems that the current management of the college — which is now in the hands of a clique of Franco-American religious who have no other ambition but to turn Assumption into a large[,] large[,] large college, wishing in that way to ape the College of the Holy Cross, more than a century old — has committed itself to the path of anglomania and anglicization to the death[,] for purposes that, from my point of view, are far from being laudable."

The editor accused the Assumptionists of "bending to the smallest wishes" of Bishop Wright; of conducting their June 7 graduation ceremonies almost entirely in English; of having an English-speaking person telephone *Le Travailleur* office for a list of French-language newspapers that might help with the reconstruction; of putting a sign outside the temporary campus at 1010 Main Street that read "Assumption College," in contrast to the sign that had for years identified the *Collège de L'Assomption* in Greendale; of turning out a special tornado issue of its magazine written almost entirely in English. "And since," he concluded, the *Collège de L'Assomption* had "molted into Assumption College and Assumption Prep, and since it oozes anglosaxonism and Irish-Americanism out of every pore of its reconstruction, its motto of fifty years, *Je maintiendrai* is nonsense and hypocrisy! My dear fathers, replace your noble *Je maintiendrai* with another more suited to your new aspirations: 'I give up'!"[27]

In a more subdued tone, in December 1953 the executive committee of the Association Canado-Américaine, headquartered in Manchester, New Hampshire, addressed a printed *"Mémoire confidentiel"* to "the authorities of Assumption College in Worcester." They noted that the association and the college had "always lived happily together, because they both followed parallel routes leading to the same goal, the survival [*survivance*] of French culture in the United States. But behold that, for reasons that do not appear very clear, you are deviating from the line laid out by your founders. If that is indeed the case, we regret to tell you that the *Association Canado-Américaine* does not approve of you."

Its disappointment stemmed from its view that the college's leaders were violating a moral obligation to conduct a college for Franco-Americans, even though the Franco-Americans had been generous in their moral and financial support. The association doubted the claim that there were not enough available Franco-American students and wondered whether enough serious effort had been made to attract them. The business courses taught in English, they suggested, would become an excuse for using English in the chapel and dining hall and at public events. The Assumptionists were right in criticizing families and parochial schools that were abandoning the use of French, but did that mean Assumption should put the last nails in the coffin of the French language? "We understand," the association said, that to be an outright Franco-American college ultimately would require "an everyday struggle, but a struggle 'through tears, sweat and blood' [*sic*] if need be, is the *'sine qua non'* condition of our survival as a people. May Assumption College get a hold of itself again, and add to its proud and loyal motto, I WILL UPHOLD, the words: TO THE END."[28]

Along with that indictment, the association sent a $1,000 check "for past services." Fr. Desautels, with the approval of Fr. Moquin, sent a response defending the leadership and fidelity of the Assumptionists. "Without casting blame on the intentions of the authors," he wrote, "we believe that the attitude expressed by your message is objectively unjust and calumnious toward us. If, in the future, the Association were to judge that what will then be the immediate past had not been a bad direction, we will be very happy to accept all the help that we need, and will [need] for a long time. I hold intact my friendship for each of the members who signed the document sent to us, and I remain convinced of their absolute good faith: it is permissible to have differences of opinion while feeling mutual esteem for one another. Accept our very strong gratitude for the services you have rendered us." In the envelope with his letter, Fr. Desautels also returned the check.[29] From Rome, Fr. Dufault pronounced the response "superb."[30]

In the changing climate of the postwar era, Fr. Desautels frequently characterized Assumption as being committed to its French heritage "from the perspective of culture rather than racial origin." Over the course of the next decade, the college would make a very public point of its continuing orientation toward French language and culture. Study of the language was encouraged through curriculum requirements, while public programs of many types were offered, usually in one of the original Salisbury Street buildings, La Maison Française, and written up in the bilingual college magazine by Fr. Polyeucte Guissard. The college also continued to seek the confidence and assistance of the Franco-American societies as it struggled to survive and develop. In September 1960, nearly seven years after the Association Canado-Américaine's *Mémoire confidentiel,* Fr. Desautels would announce from Manchester, N.H., that the Canados had pledged $100,000 toward the construction of a chapel on the campus to replace the temporary facility in the basement of Alumni Hall.[31]

In the aftermath of the tornado, the leadership of the college needed to appeal as broadly as possible for help toward recovery, and thus needed to address the public in English much more often than before. Nevertheless, the leaders of the college were committed to maintaining its French identity, both as a matter of principle and as a distinguishing characteristic that might be expected to appeal to Francophiles of all ethnic groups. To demonstrate that fidelity to tradition, and indirectly to appeal for more help, in 1954 the college marked the fiftieth anniversary of the opening of the alumnate with a "Golden Jubilee Celebration" whose theme was "France and Christendom." Since it had no suitable facilities of its own, the college centered the observances on the Hotel Sheraton in Worcester, across Franklin Street from City Hall and the Worcester Common.

The festivities, at least partially underwritten by a large group of patrons, were rich in symbolism that connected the college with France, with French Canada, and with the Franco-Americans of New England. To its festivities the college had invited scholars, church leaders, and other distinguished figures. Monday, December 6, began with an academic procession from the hotel to the nearby church of Notre Dame des Canadiens, the mother church of the Worcester Franco-Americans and the parish whose pastor had welcomed the Assumptionists to the city in 1903. The sermon was preached by the bishop of Trois-Rivières in Quebec. Presiding over the luncheon that noon was the archbishop of Boston, Richard J. Cushing, who addressed the gathering, as did the consul general of the French Republic in Boston.

The intellectual centerpiece of the celebration was a symposium, held in the afternoons and evenings of Monday and Tuesday, on the theme of France and Christendom. The audience met to hear papers — later published by the college — presented by distinguished scholars and writers. On Wednesday morning, the academic procession walked from the Sheraton to St. Paul's Cathedral for a pontifical Mass celebrated by Bishop Wright, with a sermon preached by the bishop of Springfield, whose predecessors had warily encouraged the Assumptionists prior to the establishment of the Diocese of Worcester in 1950. Lunch was followed by an academic convocation, presided over by Cardinal Francis Spellman of New York. Before an audience that included delegates representing scores of colleges, universities, and social and

Cardinal Francis Spellman accepting honorary degree from Fr. Armand Desautels, A.A., at 50th Anniversary Jubilee

cultural organizations, the college awarded honorary degrees to Cardinal Spellman; to Eugene W. Beauchamp, M.D., the first student to receive a bachelor's degree *summa cum laude* from the college and an "enlightened American citizen deeply respectful of his own cultural and ethnic traditions"; to George F. Booth, editor and publisher of the *Worcester Telegram* and *The Evening Gazette*; to Fr. Alvin Gagnon, the first Assumption graduate to be ordained a priest; and, *in absentia,* to Jacques Maritain, the renowned French Scholastic philosopher and a member of the college's advisory board, thanks to the influence of Bishop Wright. On behalf of the Assumptionists, Superior General Fr. Wilfrid Dufault, also awarded Maritain the first Emmanuel d'Alzon Medal for a life of Christian service reflecting the ideals of Fr. d'Alzon.[32] A priest who was a member of the faculty, recalling these events forty years later, remembered that "these celebrations helped encourage and reinvigorate the Assump-tionists, who had seen their work of five decades badly damaged by a 90-second tornado."[33]

The spirit of the celebration was captured by Bishop Wright in his "prefatory note" to the the proceedings of the symposium, which were published in 1956. He wondered "whether many colleges, however greater their resources, could or would have observed an anniversary in so proud and impressive a manner." France, the bishop wrote, was no longer "a political force in our corner of the world," but it remained "a luminous witness to ancient and intellectual values ultimately immeasurably more important than any political force or controls. Assumption bears that same witness. The people who built Assumption, whose sons came here, thought of themselves as bearing a witness....

Assumption perpetuates that witness; its jubilee proved that it bears its witness proudly and well."[34]

Although the college's determination to promote French language and culture persisted, the use of French in the daily operation of the institution gradually diminished, a trend reflected in the college magazine, *L'Assomption.* In April 1955, for example, the president's report was preceded by an introductory note in French explaining that the report about the state of the college's struggle for survival was in English because "the public to whom these pages are addressed includes several people who are not of French origin and do not know the language." Fr. Desautels returned to French the following month, but by 1959 his reports appeared regularly in English. In April 1956, the cover of the magazine began to carry a dual title, *Assumption* across the top, and *Assomption* printed vertically along the left margin, the two titles sharing a common "A." Beginning in the fall of 1960, the name became simply *Assumption.* Some wanted the college to shake off its bilingual past entirely. The spring 1966 issue carried a letter from a member of the Class of 1961, protesting the fact that the magazine still occasionally carried articles written in French. "When is ASSUMPTION Magazine going to be naturalized 'American'?'" the writer asked, "It just drives us all nuts to see it speaking French! We here [in Belgium] would certainly appreciate an all-American review coming from Assumption."[35] Beginning in 1967 the magazine was written entirely in English.

Evolving name of college magazine

Building the New College

In July 1954, Fr. Desautels had written a report on the college's "current situation and our immediate needs." He conveyed the consultant's cautionary message regarding a fund drive, along with a copy of Bishop Wright's letter endorsing that judgment, and wrote as if a delay of one or two years to develop "leadership" had become his policy. During that period, some solicitation would be made among special organizations, such as the Franco-American societies, which might be induced to imitate the college's longtime ally, L'Union Saint Jean-Baptiste, which was planning to donate $250,000 to build La Maison Française at the front of the campus as a French cultural center. Fr. Desautels also planned to cultivate individuals such as Joseph P. Kennedy, whom Bishop Wright had brought into the Assumption project. The former ambassador had already given $150,000 for a building to be named after his deceased son, Lt. Joseph P. Kennedy, Jr., and had indicated he would increase that gift if, as Fr. Desautels now expected, the science building cost $200,000 or more.[36]

But the core of the July report was the message, "We must build." Fr. Desautels pointed to the needs and expectations of current students as well as the need for the college to be ready for the wave of postwar babies that would begin entering college in 1960. If the college were not to be prepared by then, he wrote, "it will have missed its chance.... [W]e must build, under pain of disappearing."

But what was the best approach? No one seemed very sure. Fr. Desautels offered his superiors three plans for developing the campus, the most ambitious of which was the one on which the architects, the Nault brothers, had based their model, the plan the college had already been publicizing. Construction would begin with a large dormitory designed for 240 students in semiprivate rooms. Only 160 would originally be housed there, temporarily leaving room in the building for classrooms, administrative offices, a kitchen, and a residential area for the religious. The architects were estimating the cost at $1,408,000, which Fr. Desautels judged could only be financed through borrowing.

The Plan B offered by Fr. Desautels would have had Fr. Louis-Robert Brassard, the Assumptionist who was supervising the building of the new convent in Greendale, take on the role of contractor at the college and supervise the construction of a number of small, simple, and inexpensive buildings, with room at the start for fewer than 100 freshmen and sophomores, while the upperclassmen continued to live in private homes. Plan C was to build only one building at a time, when the money was available, rather than resort to borrowing. It would begin with a small dormitory, while the religious stayed in Greendale and classes continued to be offered at 1010 Main Street.[37]

Just nine days after sending off that report, Fr. Desautels recorded that the house council, apparently opting for Plan B, had decided to build "the new College by small buildings in order not to put ourselves too much into debt, trying to limit the debt to a half million." He added that this had been approved by the provincial superior, Fr. Henri Moquin. But plans changed quickly and often. Five days later, the council elaborated plans for a six-building start, beginning with a dormitory for one hundred students (with a temporary dining hall in the basement), followed by a house for the religious (perhaps containing administrative offices and a chapel), a classroom building (perhaps containing a temporary library), a laboratory building, a convent for the sisters who would do the cooking, and the Maison Française, if l'Union Saint Jean-Baptiste followed through on its plans to fund the building. Five days after that, on July 22, 1954, it was decided to reduce the initial buildings to three, a dormitory for ninety students and fourteen religious, with a chapel and a dining hall; a building with classrooms, the library, and administrative offices; and a convent for the nuns. This enterprise was to be executed with Fr. Louis-Robert Brassard acting as contractor. The indecision in Worcester may explain the entry made six months later in the house council minutes by Fr. Desautels: "*Construction* Fr. Provincial will take full responsibility."[38]

Fr. Moquin apparently overrode the plans being made in Worcester. When he sought permission in May 1955 to borrow $1.5 million ($1.27 million plus the $230,000 he had been authorized to borrow in 1952), it was to build a campus modeled on Plan A. Fr. Desautels' hope for a September 1955 opening of the "new college" had by then been dashed, but permission to borrow, followed by

Original Assumptionist community at the "new college." Front, left to right: Frs. Aimé Deschamps, Antonio Laberge, Vincent Dolbec, Armand Desautels, Polyeucte Guissard, Richard Richards, Bro. Arnold Glazier. Back: Fr. Anselm Morrison, Bro. Robert Lemay, Fr. Norman Meiklejohn, Bro. Armand Lemaire, Frs. Alexis Babineau, Louis Dion, Ernest Fortin, Bros. Jon Poehler, Odoric Roy, Fr. James Doyer

ground-breaking in the summer of 1955, allowed him to move his target to September 1956. Unfortunately, the devil that had historically frustrated such expectations at Assumption struck again. In the December 1956 issue of the college magazine, Fr. Desautels sent a message from his long-sought office at 500 Salisbury Street. "We had at first thought we would occupy these premises in September. Alas! I, too, had believed it. I had believed first in the date of August 15, then of September 15... and of October 8... and of October 15... and of October 21. After that I stopped believing in it and it was then, on November 3, 1956, that the students occupied their new dormitory. The community rejoined them on November 9." The original religious community at the new college campus was made up of 12 priests and five lay brothers. Remaining at "the Prep" were 28 Assumptionists, including 15 priests, 12 lay brothers, and 1 choir brother.[39]

The new campus, when its occupants moved in, was barely in presentable condition. A visitor, Fr. Desautels wrote, would need to wear boots to avoid being covered with mud up to the ankles. There were still men at work in the dormitory occupied by both students and religious. One could visit La Maison Française and the library attached to it, "but it lacks so many things!" And "if you set foot in the laboratory building dedicated to the memory of Lieutenant Joseph P. Kennedy, Jr., you would find little but walls."[40] In one state of completion or another, there were five buildings on the campus. The dormitory, later to be known as Alumni Hall, was home to students and the religious community, with a chapel in the basement. The dining hall, later to be named for a benefactor, Mae M.C. Taylor, was being served by sisters from Holland, the Oblates of the Assumption, founded by Fr. Emmanuel d'Alzon nearly a century earlier. (While awaiting a convent on the new campus, the sisters were transported more than four miles from Greendale each morning and back in the evening.) The Kennedy Science Building was under construction, as was La Maison Française. The fifth structure was a central heating plant that served the other buildings. Although the accommodations were barely adequate, the decades-old dream that Assumption College might occupy a campus of its own, away from the high school, had come true.

Growing into the New Campus

From November 1956, when students and religious moved into their unfinished dormitory, until the end of his presidency in 1964, Fr. Desautels and his staff were engaged in a constant struggle to increase enrollment and to construct the necessary facilities. On the day the religious community moved to the new campus, Fr. Desautels reported that he had been in touch with a consulting firm that would advise him on the fund-raising campaign that had been postponed until the appropriate moment. In February 1957, he and the consultants were thinking about a plan to raise $5 million to $6 million over the course of the next ten years. To Fr. Dufault in Rome, he wrote, "We have entrusted everything to Fr. d'Alzon, and a Mass is said in this intention every day."[41]

In the spring of 1957, Fr. Desautels listed the most urgent building needs: a convent for the Oblate sisters who were still commuting from Greendale, a second dormitory, and a gymnasium. In the near future, he projected, the school would also need a new building, proposed to be built in the shape of a Y, one wing to provide administrative offices, another classrooms, and the third a residence for the religious community. The financial resources did not materialize quickly. Two years later, after the Oblates had threatened to return to Holland if they were not given the promised convent, the earth had been broken for its construction, which was being financed by a substantial pledge from Bishop Wright. The president was still hoping to begin construction within the year on the second dormitory and the gymnasium. Space for boarding students was so desperately needed that plans were under way to put forty new students in the Greendale convent — after the Oblates had moved to Salisbury Street — and bus them back and forth. As for the gymnasium, because available funds were practically nonexistent, tentative plans were to begin with a simple "enclosed playing area" that could later be improved into "a beautiful gymnasium." Construction of the second dormitory — eventually named after Fr. Desautels — began in November 1959, and the convent — named after Bishop Wright — was finally occupied by the nuns that Christmas.[42]

Nearly two years later, in 1961, Fr. Desautels continued to lament the lack of a gymnasium on the campus and listed another urgent need: "We are becoming a College without adequate classroom facilities. Are we to build more, or stop the progress?" The gymnasium opened in 1963, nearly ten years after the tornado, seven years after the arrival of students on the new campus. Classroom space was so deficient that the college rented rooms from neighboring Notre Dame Academy, and rooms below the main arena of the gymnasium were used for classes. In 1963, construction began on an annex to the Kennedy Building, designed to provide classrooms, laboratories, and faculty office space by the fall of 1964.[43] By that time, the student body had grown to 420 members, and Fr. Desautels' twelve-year presidency had come to an end.

Welcoming Worcester, Welcoming Change

The pioneer president of the new college was widely recognized as an inspired and tireless leader in the struggle for survival and a new identity after the tornado and during the school's subsequent development. His presidency was indeed marked by efforts at opening the college to the community that surrounded it, to innovative educational programs, and to members of other faiths.

An early example of educational innovation was the "evening school" he promoted, first as dean, then as president. Taking advantage of the special skills of the faculty members, in 1951 the college began offering for Worcester-area adults noncredit courses in French and Russian. By the fall of 1952, sixty students were enrolled on French, fifteen in Russian, and over twenty in Gregorian chant. By then, Fr. Desautels had begun a program he conducted himself, an evening course on religion "for Catholics and non-Catholics," at no charge, with no registration, and even "without names." By October 1952, he had more than eighty people making their way to his classroom once a week.[44] These were also the years Fr. Desautels was leading the faculty discussions on the changing nature of the college's French identity and the implications of that change for admissions and curricula in the high school and the college.

After the tornado, cultivating good relations with the local community became a matter of financial strategy as well as educational opportunity. In April 1955, Fr. Desautels could describe a variety of "extension courses" the college was offering: The Evening College provided an opportunity — on the Assumption Prep campus — for students to take courses in French, Russian,

Spanish, Italian, German, Polish, English, and philosophy, leading, if desired, to an associate of arts degree. The adult education program offered, "simply as a service to the community," noncredit courses on religion and "special courses" on the family. Teaching a regular Evening College course, a new faculty member, Dr. Nicholas T. Joost, Jr., was lecturing on American poetry each week on Channel 14, WWOR-TV. The six-week summer school in French was about to confer its first master of arts degree.[45]

By August 1955, the president was announcing the establishment of a new educational center for adults, offering a set of eight-week, noncredit courses designed for workers in Worcester industries. The first courses, to be offered in January, included How to Prepare for Retirement, Photography for Beginners, Planning and Financing a Home, How to Get Ahead on Your Job, and Supervisory Development. The courses could be taught at Assumption Prep or at the factory itself if a company was so inclined.[46]

Fr. Desautels was quite candid about the financial help the college hoped to derive from such a program. In the college magazine, he expressed confidence that it would "attract to us a great sympathy from the industrialists of the city[,] since this will give us an occasion to really render them a service. At the same time, it seems to me that we couldn't do better than to attract to Assumption by every means the sympathy of local industry. To the extent that this program succeeds, it is thus certain to help the college and will be a powerful means for us to construct the new college, which is rapidly moving forward on Salisbury Street." When the adult education program opened in January 1956 the turnout was so great that the five courses originally planned had to be supplemented by five more. In the fall of 1960, the college also began offering a part-time bachelor of arts degree for adults through its new St. Augustine Institute.[47]

Another avenue to community support opened with the decision to add, in 1957, a new concentration in Foreign Affairs as a "natural outgrowth of the special emphasis that Assumption has always given to foreign languages." The Foreign Affairs Lecture

Series, attached to the program in 1959, brought to the college a succession of important figures, many of whom made appearances elsewhere in the city during their visits. Among the speakers during the first four years were historian Samuel Flagg Bemis; radio commentator H.V. Kaltenborn; retired army General James M. Gavin; Coca-Cola chairman and former Postmaster General James A. Farley; Francis B. Sayre, Jr., dean of the Washington Cathedral and chairman of the U.S. Committee on World Refugee Problems; motion picture producer and director Frank Capra; former ambassador to Moscow Charles E. Bohlen; and Air Force Chief of Staff Gen. Curtis Lemay. Many of those talks were followed by receptions at which refreshments were served by local women who had been organized in 1959 into the Assumption College Volunteers, to be "coordinators between what the College offers and what the community needs."[48]

Gen. Curtis E. Lemay and Fr. Armand Desautels, A.A.

The openness that came to characterize the "new college" was especially evident in matters of interfaith relations, years before the opening of the Second Vatican Council in 1962. Fr. Ernest Fortin began in 1958 to host an annual gathering of Catholic and non-Catholic philosophers, mostly from the New England area "to discuss problems of common or current interest in an informal and intellectually hospitable atmosphere." In 1959, the college introduced an annual one-day Institute on the Person and the Common Good, designed to "bring together Jewish, Catholic, and Protestant leaders, and friendly disposed agnostics or others... for a discussion of the 'common good.' " At the first gathering of the Institute on May 9, 1959, around seventy-five "lawyers, doctors, heads of business firms, presidents of large industries, advertising executives, religious leaders of all faiths," and academics participated.[49] When the sixth annual meeting was held in 1964, an official from the Anti-Defamation League of B'nai B'rith declared that the institute had "served a national role because it was the predecessor of 30 large conferences on inter-faith relations in the United States." In 1964, during the final year of Fr. Desautels' presidency, the Brotherhood of the nearby Congregation Beth Israel presented him with its Good Neighbor Award. One speaker remarked, "His warmth and brotherliness prefigured the attitudes and actions of the late Pope John, whose words to a Jewish delegation, 'I am Joseph, your brother,' found a living model in Father Armand long before they were uttered."[50]

The superior general, Fr. Wilfrid Dufault, had remarked early in 1960 on the development of new ties between the college and the city after the tornado. "A new spirit arose," he wrote, "thanks to the zealous initiatives of the college staff, and to the understanding and generous cooperation of scores of friends in the various spheres of activity in the community."[51] Fr. Louis Dion, a Worcester native and the next president of the college, praised Fr. Desautels as the man who "brought Worcester to Assumption College and Assumption College to Worcester."[52]

Notes

ABBREVIATIONS

AC — Archives of Assumption College, Worcester, MA

AG — Archives of the Assumptionist General House, Rome

AP — Archives of the Assumptionist Provincial House, Brighton. MA

PA — Archives of the Assumptionist Province of Paris

SP — Archives of the Diocese of Springfield, MA

1. Dufault, "Rapport semestriel," Jan. 11, 1947, AP; "Archives de la maison,"III, 54, AP.
2. "Archives de la maison,"III, 57–58, AP; House Council Minutes, 24–25, AP; Dufault, "Rapport semestriel," AP.
3. *L'Assomption* 9, 7 (Feb. 1947): 145–47; *L'Assomption* 9, 8 (April 1947): 169–70; Moquin and Richards, *Assumptionists in the United States,* 79.
4. Desautels to Quénard, June 1, 1947, AG 2PL-322.
5. *L'Assomption* 9, 11 (Oct. 1947): 223–25; 9, 12 (Dec. 1947): 253–55; Moquin, Memoir, (1988), AC; "Archives de la maison,"III, 75, AP.
6. House Council Minutes, Feb. 6, 1947, AP; Moquin, "Rapport annuel," Oct. 23, 1947, July 15, 1950, AP.
7. Moquin, "Rapport semestriel," Jan. 3, 1948, AP.
8. Moquin, "Rapport annuel," July 12, 1949, PA; "Rapport annuel," July 10, 1951, AP.
9. *L'Assomption* 13, 2 (March–April 1952): 32.
10. Moquin and Richards, *Assumptionists in the United States,* 78.
11. Desautels, "Rapport semestriel," Jan. 14, 1953, PA; House Council Minutes, 40–41, AP; *L'Assomption* 9, 4 (Aug. 1946): 92.
12. House Council Minutes, 46–49, AP (emphasis in original).
13. *L'Assomption* 11, 6 (Dec. 1950): 113–14; Desautels, "Notre Orientation," enclosed with Desautels to Dufault, Oct. 21, 1952, AG 2PK-137.
14. *L'Assomption* 14, 1 (Jan.–Feb. 1953): 2–3.
15. *L'Assomption* 14, 2 (March–April 1953): 28–30; Assumption College Bulletin, 1953–54 (Worcester, 1953), 17–22.
16. Moquin, Memoir, AC.

17. "Lettre du P. Polyeucte Guissard au T.R. Père général," *Lettre à la Famille,* N.S. 155 (Sept. 1, 1953).
18. "Archives de la maison," III, 204, AP.
19. *L'Assomption* 14, 8 (Sept.– Oct. 1953): 131–32.
20. *Assumption* 21, 1 (Winter 1960): 4; Desautels, "Rapport annuel," Aug. 12, 1953, AP.
21. [Moquin], "Memo," June 12, 1952, AG 2 PK-133; Moquin, "Rapport annuel," July 15, 1950, AP.
22. [Moquin], "Memo," June 12, 1952, AG 2 PK-133; Superior General [Wilfrid J. Dufault], "Worcester Loan," June 19, 1952, AG 2PK-116.
23. House Council Minutes, 7, AP; Desautels to Dufault, Sept. 24, 1953, AG 2PK-154; *L'Assomption* 15-2 (Feb.–March 1954): 35.
24. Desautels to Dufault, Sept. 24, 1953, AG 2PK-154; Desautels, "Rapport sur la situation actuelle du Collège et nos besoins immédiats," July 1954, AG 2PK-169. The House Council also signed this report on July 3, 1954.
25. House Council Minutes, 71, AP; Desautels, "Rapport annuel," July 1, 1954, AP.
26. Wright to Desautels, June 21, 1954, AG 2PK-170.
27. Wilfrid Beaulieu, *Le Travailleur,* Oct. 1, 1953.
28. Adolphe Robert et. al., "*Mémoire confidentiel,*" Dec. 23, 1953, AC.
29. Desautels to Dufault, Feb. 22, 1954, AG 2PK-158.
30. Dufault to Desautels, March 19, 1954, AG 2PK-138.
31. Desautels, "Rapport annuel," Aug. 12, 1953, AP; *Assumption* 31, 3 (May 1970): 14.
32. "Assumption College Golden Jubilee Celebration," AC.
33. Richard Richards in Moquin and Richards, *Assumptionists in the United States,* 80.
34. *France and Christianity: Golden Jubilee Symposium 1954, Assumption College, Worcester, Massachusetts* (Worcester, 1955), 1–2.
35. *Assumption* 27, 2 (Spring 1966): 42.
36. Desautels, "Rapport sur la situation actuelle du Collège et nos besoins immédiats," July 1954, AG 2PK-169; *L'Assomption* 16, 2 (April 1955): 66; Desautels to Dufault, Sept. 17, 1956, AP.
37. Desautels, "Rapport sur la situation actuelle du Collège et nos besoins immédiats," July 1954, AG 2PK-169.

38. House Council Minutes, 86–89, 93, AP.

39. Desautels to Dufault, May 26, 1955, AG 2PM-33; "The New Assumption College," fundraising brochure [1955?, 3], AC; *L'Assomption,* 17, 6 (Dec. 1956): 208; *L'Assomption* 17, 5 (Oct. 1956): 182–83.

40. *L'Assomption* 17, 6 (Dec. 1956): 209.

41. Desautels to Moquin, Nov. 9, 1956, AP; Desautels to Dufault, Feb. 27, 1957, AG 2PM-53.

42. *Assomption/Assumption* 18, 2 (Spring 1957): 266–67; Desautels to Moquin, Nov. n.d., 1958, AP; *Assomption/Assumption* 20, 2 (April 1957): 3; *Assomption/Assumption* 20, 2 (Spring 1959): 3; *Assomption/Assumption* 21, 2 (Spring 1960): 6.

43. *Assumption* 22, 2 (Spring 1961): 3; *Assumption* 24, 4 (Fall 1963): 3; *Assumption* 24, 4 (Fall, 1963): 8.

44. *L'Assomption* 11, 11 (Sept.–Oct. 1951): 122; Desautels to Dufault, Oct. 21, 1952, AG 2PK-137.

45. *Assomption/Assumption* 21, 3 (Summer 1960): 17–18; *L'Assomption* 16, 2 (April 1955): 70.

46. *L'Assomption* 16, 4 (Aug. 1955): 126–27.

47. *L'Assomption* 16, 4 (Aug. 1955): 127; Desautels, "Rapport semestriel," Jan. 24, 1956. AP; *Assumption* 21, 4 (Fall 1960): 3.

48. *Assomption/Assumption* 18, 4 (Aug. 1957): 367; *Assumption* 24, 2 (Spring 1963): 7–8; *Assomption/Assumption* 20, 1 (Winter 1959): 23.

49. *Assomption/Assumption* 22, 1 (Winter 1961): 5; *Assomption/Assumption* 20, 2 (Spring 1959): 6; *Assomption/Assumption* 20, 3 (Summer 1959): 15–16.

50. *Assumption* 26, 1 (Winter 1965): 9; *Assumption* 25, 2 (Spring 1964): 5.

51. *Assumption* 21, 1 (Winter 1960): 4.

52. *Assumption* 26, 2 (Spring 1965): 3.

Chapter Five
From Confusion to Crisis (1964–1972)

In 1964 Fr. Wilfrid Dufault was elected to another term as superior general in Rome. This time he called Fr. Henri Moquin to join him as treasurer general. Fr. Armand Desautels, after twelve years as superior and president, was elevated to the office of provincial superior, which took him to New York City. With these appointments, the three early leaders of the Franco-American Assumptionists would, for the first time since the 1940s, all be absent from administration of the college. Despite all the Franco-American pioneers had faced and surmounted, their successors, mostly younger men, would find it even more difficult to lead the new college through a decade of social and political turmoil accompanied by severe financial dangers.

At the request of Fr. Moquin, Fr. Louis Dion agreed to become the president of the college. Fr. Dion had served as professor, dean, admissions director, and registrar, and, from 1959 to 1961, he had filled the Assumptionist position in Moscow as apostolic administrator and minister to Catholics in the diplomatic corps. The appointment of Fr. Dion was accompanied by a major structural change.

For the first time, someone other than the college president was chosen to serve as superior of the religious community. Fr. Desautels had been advocating this arrangement for at least a decade, arguing that the extensive travel required of the president did not leave adequate time for the work of supervising and guiding the religious community. The new superior was Fr. Ernest L. Fortin, who also served as vice president of the college. Fr. Fortin was widely admired as a theologian and political philosopher, as an exceptionally productive scholar, and as an inspirational teacher.[1]

In separating the superiorate from the presidency the Assumptionists were inadvertently creating a situation that could lead to confusion of authority. In the past, matters on the agenda of the college had automatically been on the agenda of the religious community as well, and what affected the religious community intimately affected the college. The superior had been clearly in charge, and only the need to conform to American terminology had led the institution to confer on the superior the internally superfluous title of president. Now the two positions were to be held by different men, but it would not be a simple matter to disentwine the agenda of the college and its president from that of the religious community and its superior. In the administrative hierarchy of the college, Fr. Fortin and the subprior, Fr. Alexis Babineau, worked under Fr. Dion. In the religious community, however, Fr. Dion was outranked by both of those administrative subordinates.

In the formal organization of the college, the top level was occupied by the Board of Trustees, made up of the provincial superior and his two counselors. Next came the Board of Directors, consisting of the same three men plus the provincial treasurer, the superior of the college community, and the college president.[2] Lines of authority at the college were thus further complicated by the presence of administrators other than the president on the Board of Directors. Starting in 1965, in addition to Fr. Dion, Fr. Fortin sat on the board as his peer, as did two

Fr. Louis Dion, A.A.

Fr. Ernest Fortin, A.A.

Fr. Alexis Babineau, A.A.

other men who were Fr. Dion's subordinates in administration, the treasurer, Fr. James R. Doyer, and the assistant treasurer, Fr. Louis-Robert Brassard.[3] Fr. Desautels, when he served as both local superior and president, had spent a dozen years reporting principally to his provincial superior, Fr. Moquin. They had conferred extensively on most matters of college administration as well as on issues involving the religious community itself. In his new position as provincial superior, Fr. Desautels continued the practice of communicating on both community issues and college administrative issues with the local superior — who in this new arrangement was Fr. Ernest Fortin, the vice president rather than the president of the college.[4]

Following the Founder

Years later, Fr. Dion recalled that he had "found administration very difficult." His time was largely consumed by the need to raise funds. When members of the lay faculty complained that the organization of the college and the distribution of authority within it were not clear, it was the vice president, Fr. Fortin, who took the lead in producing a new organizational chart. The lines of authority showed the president consulting with the outside Advisory Board and an internal College Council. Below that level, administrative responsibilities were divided into three branches. As vice president, Fr. Fortin was responsible for academic affairs, including "noncurricular" areas such as the dean of men's office and the athletic office. Fr. James Doyer would be in charge of business affairs, and another Assumptionist, Fr. Alexis Babineau, A.A., would be responsible for "public affairs."[5]

Fr. Dion inherited the presidency at a time of unusual turmoil in American society. The postwar "baby boomers" were arriving in unprecedented numbers on college campuses, many of them ready to ask questions about the ways institutions were being conducted. Confidence in authority, especially among the young, was eroding as a result of many factors, prominently including the Civil Rights Movement and, by 1965, growing opposition to the Vietnam War. For Catholics, new possibilities were also being opened up by the Second Vatican Council. These winds of change, along with many others, contributed to unrest in the Assumption community, where some uneasiness had already been provoked by a rapid increase in the size of the student body, accompanied by a dramatic rise in the proportion of lay persons on the faculty.

During the college's first year (1956–57) on Salisbury Street, 201 students had enrolled. When Fr. Dion began his presidency in 1964, there were 420. The following September, thanks in part to the opening of a third dormitory (later named Founders Hall),[6] there were 507. By the fall of 1968, there would be 699. Similarly, when the college opened in 1953 following the tornado, two full-time lay professors (Lucien H. Desjardins and Raymond J. Marion) were listed in the catalog along with 15 Assumptionists. In 1959, for the first time, the number of lay people (16) exceeded the number of Assumptionists (13). When Fr. Dion took office — only 11 years after the tornado — 34 laymen and 1 lay woman made up 74 percent of the faculty of 46.

Traditional ways of dealing with students seemed increasingly out of date. As recently as 1952, a student who insisted he could not be forced to attend morning Mass had been admonished and told he would be put on probation if he persisted. But even before the tornado, some rules were relaxed. In 1953, Fr. Desautels and his council had decided to allow the college-level students "certain liberties" in an attempt to quiet the "malaise" among them. They would be allowed to use radios, to choose whether to attend meals, evening prayers, and Benediction of the Blessed Sacrament, and they would be free to leave the campus without asking permission between or after classes, as long as they returned by 8:00 p.m.[7]

Fr. Norman A. Meiklejohn, when he served as dean of men from 1956 to 1960, found that the "students would not hear of

Fr. Norman Meiklejohn, A.A.

observing rules that we had unquestioningly observed in 1946, or even in 1953." He presided over a relaxation of the campus rules and an expansion of social activities that succeeded, he believed, in winning "the good will and cooperation of the students." After a six-year absence, he returned to the college in 1966 and took up his former post, under the new title of dean of students. "Once again," he reported, "I find that rapidly changing times have changed our students." They came from a broader range of states and countries; many had been "raised in affluence"; they had been marked by "Vatican Council II, peaceful demonstrations for civil rights, [and] concern and protest over the U.S. Government's involvement in the war in Vietnam." The much larger lay faculty, he wrote, "with degrees from top-level universities... stimulate and inspire the students in a manner unprecedented in the college's history." The students were even "more refractory to rules than ten years ago," wanted "a good deal more independence," and wanted "their opinions to be considered before educational policies are formulated." Fr. Meiklejohn pledged to abide by "the principle that educating men for tomorrow cannot be done by adhering to the policies of yesterday."[8]

In 1954, as part of the fiftieth anniversary festivities, the college had adopted a new motto. Leaving the French-oriented *Je maintiendrai* to the prep school, the designer of the new coat of arms, Bro. Gerard Brassard, selected a faith-centered substitute: *Donec formetur Christus in vobis* (Until Christ be formed in you), a favorite passage of Fr. d'Alzon's, drawn from St. Paul's Epistle to the Galatians, 4:19. Some of the early expressions of dissent at the new Assumption also involved issues related to religion. Fr. Desautels in 1963 expresses his worries "about problems of faith among too large a number [of students]. There are some who openly profess not to believe, and several deliberately miss Mass on

Sundays." The stirrings of discontent were exposed with considerable candor in the college magazine edited by Fr. Aimé Deschamps. In the summer of 1965, he reproduced an article from the student newspaper, *Le Provocateur,* a mimeographed publication that had begun to appear in 1964. Rick Ryscavage of the Class of 1967 asked, "What causes the discontent with the Church and Catholic education, which so frequently comes up in the students' conversation?" He put most of the blame on theology courses, charging that it was "common knowledge among Catholic students that theology courses are the easiest, dullest, and most disconnected parts of the curricula." He called upon priests teaching theology to explore with their students controversial approaches such as that of Pierre Teilhard de Chardin. Unless they did, he argued, they would be "defeating the whole 'aggiornamento' purpose of Vatican II. Once again, the Church will fall dangerously behind."[9]

This contribution led to a regular space in the magazine for Mr. Ryscavage to present an "undergraduate slant." In the fall of 1965, he discussed the "split" between generations. "There was a time," he wrote, "when the older generation consistently triggered the ambitions and values of the young. No longer is this true. Contemporary youth find their heroes within the ranks of the young, and they have been known to scoff at the 'stained glass voices' of their elders." In the next issue he asked, "Point Blank: just what are the long range goals of Assumption College? As a student here, I seriously question whether any exist. It is hard to speak of experimentation and imagination, scholarship and tradition, in an aimless context. Without an overriding purpose, we drift in a way that is more hit-or-miss than pragmatic. The sooner the college defines itself, the easier it will be for the students to lose their sense of injured merit. Young people often reflect their educational environment which, at the present time[,] is both confused and confusing." Also in 1966, another student wrote about "being on campus in the midst of change." Pointing to frequent modifications in the curricular requirements, he complained that students could not tell "where all this is leading to" and speculated that "perhaps there has not been enough of an attempt to define the aims in a clear manner."[10]

As if in response to such criticisms, the 1966–67 college catalog appeared in two editions. It is instructive to compare the two. Under "Aims and Objectives," the first version referred to an enthusiastic faculty, small classes, the development of critical thinking, preparation for graduate and professional education, and extracurricular activities. The revised version, under "Aims," spent three pages elaborating on three declarations: "Assumption is a Catholic College," "Assumption is a College of Liberal Arts and Sciences," and "Assumption is a College of French cultural traditions." In each case, the text emphasized that these attributes, far from being products of changing times, were the essential components of the college's tradition.[11]

However useful such reaffirmations may have been, individuals who were anxious about changes in the college may not have felt comforted by several articles that appeared together in the fall 1967 issue of the college magazine. Fr. Louis Dion wrote about the Assumptionist decision, announced by Fr. Desautels the previous January, to replace the Board of Trustees, made up of five Assumptionists, with a new board, two-thirds of whose members would be lay people. The goals were to bring in business expertise at the decision-making level, to assure major donors of "the most effective use of funds," and "to assure the placement of the diminishing proportion of Assumptionists...in types of work more directly linked with their Priesthood."[12]

Dr. Oscar E. Remick wrote about the new Ecumenical Institute of Religious Studies, scheduled to begin offering courses in January 1968. The program was designed to serve the educational needs of religious educators and members of the clergy in Orthodox, Protestant, and Catholic parishes and schools. The institute had been developed by Fr. Ernest Fortin and by Dr. Remick, who, after serving three years as pastor of the First Congregational Church of Paxton, Massachusetts, had joined the Assumption faculty in 1966 as an assistant professor of philosophy and theology. The institute's first year would be subsidized by a grant of over $40,000 from the Fuller Foundation of Worcester, $25,000 of which would be used to buy library books in religion and theology. One branch of the program would offer graduate courses, leading, if desired, to the master of arts degree; the other would be a certification program open to all, but geared toward individuals involved in Christian education programs at the parish level. The founders were hoping to find the means to build "a kind of 'Ecumenical Center for Continuing Education'" with a library, classrooms, lecture halls, and seminar rooms.[13]

Dr. Oscar E. Remick

The winds of change were clearly continuing to blow. Lay trustees would be an important modification, as would collaborating on an equal footing with Protestant and Orthodox Christians on a significant educational project. As if for good measure, Fr. Fortin weighed in with an article entitled "Coeducation at Assumption." The graduate school, the St. Augustine Institute, and the summer school had been coeducational since their inceptions, and for several years student nurses from the St. Vincent Hospital School of Nursing had been taking some of their courses on the Assumption campus. However, the local press had recently reported that college officials were considering the admission of women to the full-time undergraduate college, and Fr. Fortin's report to the college's alumni and friends left little doubt that consideration was getting very close to commitment. Coeducation "does not ordinarily present a direct financial advantage," Fr. Fortin wrote, and "it is agreed on all hands that economic considerations do not of themselves constitute a sufficient reason for admitting women." The "real reason" was that a woman was less restricted to the home than in the past, and "what is needed today is an education that corresponds to the new demands that are being made upon her." Also, "Is not education, which is above all education for life, best carried out in a well-rounded academic community which imitates, albeit in its own original way, the normal conditions of life itself?" He reported that members of the faculty and administration were "overwhelmingly in favor" of coeducation, as were "a majority of the students who have been polled."[14]

Assumptionists, many alumni, and a few lay professors remembered an Assumption College in which priests and brothers, high school boys, and young college men had lived and worked together, in considerable isolation, in a bilingual community in Greendale. Thirteen years after the shock of the tornado, it had become more than obvious that there were many ways other than physical appearance in which the "new college" would not resemble the old.

During the remainder of his presidency, Fr. Dion searched for ways to finance buildings, including a new chapel (for which ground had been broken in June 1964). He presided over the 1966 completion of the classroom, laboratory, and office building annexed to Kennedy Memorial Science Hall, as well as the purchase of land in 1967–68 that added forty-five acres to the campus. He also witnessed the departure in 1967 of the Oblates of the Assumption, who left their cooking duties to take up social work in the Diocese of Worcester.[15] When he learned that his old post in the Soviet Union would soon be vacant, Fr. Dion urged the provincial superior, Fr. Desautels, to send him back to Moscow. The request was granted, and early in 1968 Fr. Dion submitted his resignation to the Board of Trustees, effective June 30 at the latest. He later recalled that in the news media his change of positions was sometimes interpreted as a heroic choice, trading the life of a college president in Worcester for a situation of some danger and isolation in Moscow. In fact, he said, considering the difficulties of his position at Assumption, "I was very happy to go."[16]

Teaming Fr. Bissonnette with Dr. Remick

On April 19, 1968, the trustees asked their chairman, Fr. Ernest Fortin, to "appoint a committee to find a new president." Fr. Fortin appointed himself and two other Assumptionists as well as three laymen, who were at that time attending meetings and participating in trustee business, even though the formal transfer of authority to a new lay-clerical board had not yet taken place. In June, the fully and officially constituted lay-clerical board elected Fr. Fortin its chairman, agreed that lay candidates might be considered for the presidency, and ordered the search committee to appoint an acting president. The committee appointed its chairman, Fr. Fortin, to that position, which left him simultaneously holding the titles of chairman of the board, acting president, and chairman of the presidential search committee.[17]

As early as June, possibly earlier, concern spread among some Assumptionists that the search committee might recommend as the new president of the college Fr. Fortin's partner in the Ecumenical Institute, Dr. Oscar Remick, a Congregational clergyman. In mid-June, a member of the search committee received a second-hand report that Superior General Wilfrid Dufault in Rome "has been alerted of the possibility of a non-Catholic president for the College. He declares that the thing is inconceivable and he is absolutely opposed to it." In July, Fr. Desautels, still provincial superior, threatened in a letter that if he did not receive assurance "that no non-Catholic will be considered for the position," he would write to other trustees and, if necessary, resign from the board and publicly state his reason for doing so. In a postscript he added, "In case I have not stated clearly enough WHY I am so opposed to a non-Catholic, it is because the President of a College is the image of the institution: ours is a Catholic College, engaged in ecumenical work, but a Catholic College."[18]

The search committee reported to the board on October 11, 1968. As summarized in the minutes, Fr. Fortin said deliberations

Fr. Georges L. Bissonnette, A.A.

and consultations "all pointed to the fact that it would be unwise to name a layman to the presidency of the College at this critical juncture." Given the "many important changes" made at the college recently, "another radical departure from tradition would not be well received by many alumni and friends." A lay president would also add "another high salary" that would "cripple the critical financial situation of the College." The committee recommended the appointment as president of Fr. Georges Bissonnette, a graduate of the college who held a doctorate in Russian intellectual history from Columbia University, had served in the Assumptionist post in Moscow, and had worked at the college as a faculty member and as academic dean.

Fr. Fortin added, according to the minutes, that "many members of the College community had strong feelings about the kind of person who should be named vice president and dean when Fr. Georges became president." The board agreed to discuss this matter once the new president had been appointed, then unanimously elected Fr. Bissonnette president, effective immediately. The new president joined the meeting, expressed his thanks, and recommended that, "in the spirit of ecumenicity already demonstrated in various ways by the college," Dr. Remick be appointed vice president and academic dean.

One trustee raised a point of order, noting that no such item had appeared on the agenda for the meeting. Another warned against a "hasty decision," noting that the Faculty Evaluation Committee had not endorsed Dr. Remick's appointment as dean. Others responded by saying that, although the faculty had not been polled, many faculty members had been involved in the deliberations "and that to postpone the appointment of a Dean would perhaps complicate matters and create new difficulties at this time." After a discussion about the appointive powers of the trustees and the president, the board appointed Dr. Remick vice president with no negative votes and one abstention. Fr. Bissonnette then notified the board that he was appointing Dr. Remick academic dean.[19] The appointment of an administration led by a Catholic priest and a Protestant minister was widely hailed as an expression of the ecumenical spirit of Vatican II and of the college.[20]

If Fr. Bissonnette and Dr. Remick inherited the discontents that were troubling the nation and the college in the late 1960s, they also inherited the spirit of innovation that had become a fact of institutional life. Even before they took office, in the fall of 1968, the Institute for Social and Rehabilitation Services was founded, "joining the Ecumenical Institute of Religious Studies as another autonomous unit within the structure of the College."[21] The institute, led by Dr. George S. Elias, offered both graduate and undergraduate programs to students interested in serving disadvantaged and handicapped persons; its program would become a popular major among undergraduates.

Coeducation

The new lay-clerical board of trustees had decided that the college, if it was to be economically viable, needed to grow rapidly to a student population of twelve hundred. To promote that expansion, they had decided in April and more formally in June 1968 that the college would open a coordinate college for women in September 1969. The coordinate college was to have its own dean and auxiliary staff, its own rules and regulations in nonacademic areas, and its own opportunities for student leadership. In the 1969 freshman class of 324, 124 were women; the size of the student body rose from 699 students to 864. By early 1971, the administration and trustees would decide, due to financial constraints, to eliminate the administrative costs of the coordinate college by merging the first two classes of women, and their successors, into an unequivocally coeducational college.[22]

Also during the first year of the new administration, Dr. Remick announced a new set of curricular requirements. The old

Welcoming the pioneer women

integration-concentration model, inaugurated the year of the tornado, had given way over the years to a growing number of majors and a recurring revision of what would constitute "core" courses that all or most students would take in common. Student and faculty calls for greater student choice were among the major factors that produced the 1968 curriculum. To meet general education requirements, students were to select two courses in seven out of eight "distribution" areas (English, a foreign language or literature, history, mathematics, natural sciences, philosophy, religious studies, social/behavioral sciences). Students would also complete requirements for a major, leaving them fourteen to twenty elective courses. They were urged to consult with faculty advisors in the process of making their choices.[23]

Also from the first months of the new administration, construction was under way to house the new women students. Unlike the three previous dormitories with their long corridors, the three new buildings, designed for fifty students each, were clustered on high, sloping terrain to the northeast of the heating plant and Kennedy Hall. Initially two of the new buildings were to be occupied by women, the third by men. Taylor Dining Hall was also to be renovated and expanded. The college's opening-day demon did not allow this opportunity to pass. Fr. Bissonnette discovered during his first summer of the job that he would not be able in the fall "to welcome the Freshman Class of 1969–70 to spanking new dormitories and a completely remodeled and enlarged dining hall." Instead, his "most urgent problem [was] to find temporary quarters— off campus — for more than a hundred returning upperclassmen [to make room on campus for the women] while trying to feed about 180 more students than last year in a dining hall whose walls have not become elastic." He blamed the problem on "the state of the national economy" and on the demands of labor unions. Rather than move upperclassmen off campus, the college turned some of the two-man rooms in the large dormitories into temporary "triples," making space for the women to move into Alumni Hall. The pioneer class of women was welcomed onto the campus on September 12, 1969, each member receiving a rose from the men who had lined up to help carry their luggage and other belongings. When the first of the "Hill" dormitories was ready for occupancy in November, the fifty seniors (all males) who had been deprived of their anticipated space on the hill laid claim to the building. A college committee decided in favor of the women, after they volunteered to "triple" temporarily in that new facility, thus vacating Alumni and allowing the men to "untriple at once."[24]

Assumptionist Anxieties

Housing had become an increasingly pressing issue for others on the campus as well. The Assumptionist fathers and brothers had been living in student dormitories since 1956, a noisy environment that many did not find conducive to community religious life or to spiritual and intellectual growth. College authorities had repeatedly placed higher priorities on residential and athletic facilities for students than on providing a proper residence for the religious. In 1967, Fr. Armand Desautels as provincial superior admonished the local superior, "While waiting for the faculty residence, it is imperative that the religious be in a place where they will have no students above them — this cannot wait another year." Later that year, the Assumptionists at the college held a series of meetings to discuss the matter of a religious residence. Hanging over the deliberations was the fact that the college was operating at a deficit, estimated by one of the participants at $200,000 for 1966–67. Fr. Joseph Pelletier pleaded the urgent need for action. Fr. Ernest Fortin, the local superior and college vice president, while agreeing on the need, did not consider the matter so urgent. He stressed instead the need for the college to continue growing so that it could attract and keep good professors. He added that the whole matter should be referred to the Board of Trustees, which was already functioning with some lay members, even though the formal transfer of property and power would not, as things would turn out, occur until June 1968. A week later, Fr. Georges Bissonnette, then the academic dean, argued the pressing need for additional classroom and office space. When talk turned to potential designs of a residence, opinions about how the building would be used were also varied. Among the open questions, one priest said, was "How the religious community wishes to exercise its apostolate."[25]

The fact that this was an open question indicates that the continued participation of some or all of the Assumptionists in the college enterprise had become uncertain. At a third meeting on the issue of a residence, the community discussed an article written by a lay faculty member, Dr. Joseph Meier, scheduled to be published in that spring's issue of the college magazine, *Assumption.* At recent faculty meetings Dr. Meier had heard leading Assumptionists describe the college "as a base of operations for the Fathers in their apostolic work with young Catholics. Liberal arts education was presented as a means, not as Assumption's proper goal."

"In both meetings I could not help but gain the impression that the lay faculty was served with a special message, with a reminder that Assumptionists could only be interested in a college that served their good purposes, that their priesthood was too costly for liberal arts education which, after all, could be carried on by laymen, freeing the Fathers to follow the urgent call of priestly duty somewhere else. If Assumption is to stay an Assumptionist college, it must not be 'secularized.'"

Dr. Meier argued that neither the liberal arts nor the lay faculty automatically implied secularization. "The liberal arts *have* a deep religious dimension," he insisted. "Culture, art, truth are all part of religious life, not its lure." He warned that, "the presence of religious personnel [in a college] may endanger the religious dimensions of the educational encounter if the clergy exaggerates their claim to be guardians and mediators of the holy. Because secular then may come to mean non-clerical, with the implication what whatever the clergy does be considered religious." Dr. Meier urged that the Assumptionist community give up their effort to "man the administration" and concentrate instead on two essential functions: "1. As a religious community it must show what religion can do to people. This calls for an intense religious life that at the same time is open to all members of the college at all times. 2. As a community of religious scholars, it should afford, in addition to religious practice, a collateral study into the phenomenon of faith." This brought him to the matter of the college's physical facilities, which he saw as inadequate "especially for the fostering of religious community life." He therefore suggested the construction, at a secluded and quiet place on the campus, of a "center for religion" with ample space, including meeting space, for each priest; with facilities for persons wishing temporarily to share in the residential life of the community, with rooms for scholars "of all faiths and none." He "implored" the Assumptionist community "to resist the temptation of educational politics and to turn all their energies to the religious life of the community."[26]

Dr. Joseph Meier

When the Assumptionists met to discuss their residential needs in the light of the Meier proposals, discussion showed "that the ideal proposed by Mr. Meier is attractive to some and leaves others unmoved." It was suggested that the residence might have three parts, one for the members attracted to the scholarly function described by Dr. Meier, one for those who would be working in other roles on the campus, and a third for student brothers. Discussion proved inconclusive, in part because some members did not think they could express themselves freely in an open meeting.[27]

The process of creating a residence for the Assumptionists would go on for another six years, well past the presidency of Fr. Bissonnette. Meanwhile the whole question of whether the Assumptionists should continue their work at the college or concentrate on other apostolic endeavors continued to percolate. In 1970, one priest wrote privately, "I really would like to see aired all the pros and cons of Catholic higher education, and of retaining our control of A.C., and even of the value of A.C. itself. I'm not convinced of their value."[28] Soon after, a representative of Fr. Paul Charpentier — who had succeeded Fr. Wilfrid Dufault as superior general in 1969 — spent three long days interviewing the religious at the college and found signs of spiritual hope, despite the dangerous financial condition of the college. "There are encouraging signs in the efforts of a small group of young priests charged with the spiritual animation of the college. The majority of the religious absolutely refuse to consider the distant possibility of our abandoning the college.... Finally, as everywhere, here also the young (at least some of them) are not convinced of the validity of this type of apostolic engagement."[29]

Crisis

Into this troubled sea the intended partnership of Fr. Bissonnette and Dr. Remick was launched. On May 6, 1969, Fr. Bissonnette became the first Assumption president to be installed with an inaugural ceremony. A week later, the local newspapers were reporting unrest on the campus over the removal of two laymen who had been serving as associate deans. The faculty protested the lack of proper consultation on the matter, and hundreds of students met to demand the reinstatement of the two administrators. Some participants framed the uproar in terms of uncertainty about where authority resided in the institution — in the hands of the president, or the vice president, or the religious superior, or the Board of Trustees. Students and faculty were agreed that it did not reside with them, and some were searching for ways to change that situation. The perception of ambiguous leadership was not diminished by Fr. Bissonnette's departure on a long-scheduled trip to the Far East with a prospective benefactor, leaving Dr. Remick to deal with the situation.[30]

In November 1969, after a year as vice president, Dr. Remick addressed a letter to the president, with a copy to the chairman of the board, Fr. Ernest Fortin, requesting clarification of his and the president's responsibilities, indicating he would resign if a satisfactory clarification was not forthcoming. The executive committee of the trustees met with the two administrators and decided to draw a sharp line between their areas of responsibility. On December 3, Fr. Bissonnette issued a memorandum announcing that the executive committee of the board had "decided that the Vice President be delegated ... the responsibility and authority to guide the College Community toward clearly defined goals according to declared policies." While the vice president led and administered the campus, Fr. Bissonnette's task would be to lead an upcoming capital fund campaign "with its countless time-consuming details." The completion of the campaign would not "substantially diminish the need for my continued efforts in this area."[31]

The year 1970 proved to be even more difficult for the uncomfortably linked administrative team. In mid-February, the student and faculty members of the Black Studies Committee, originally appointed by Dr. Remick, complained that the trustees had not acted on their request for scholarship money for African-American students beginning in the 1970–71 academic year. A long series of meetings and publications ensued, highlighted by the firing of the dean of admissions after he had questioned figures put forth by the administration, and by an occupation of the recently completed and not-yet-dedicated Chapel of the Holy Spirit by some members of the Black Studies Committee and their supporters. The combination meeting, sit-in, and sleep-in began on Wednesday, February 25 and extended to Saturday, February 28, provoking among other reactions, a Thursday petition signed by 200 students declaring the occupation a "sacrilege" and a written request from the Assumptionist community on Friday that the chapel be vacated in the light of the agreement of trustees to meet with representatives of the committee. On Saturday morning, according to the student newspaper, the chapel was "cleaned up, washed, and swept" by the students before they left. At Dr. Remick's request, the trustees made a one-year commitment to scholarship funds for disadvantaged students and further funding for the Black Studies Committee.[32]

Prep Closing, Campaign Failure

The disturbance brought more unwelcome publicity to the college, an experience particularly disturbing to the trustees, who on February 6 had authorized the 65th Anniversary Building Campaign and Fund. The principal objective of the drive was to raise $1.573 million "from the Assumption College Family" as the college's share of the costs of a nine-story classroom and office building and a residence for the religious community. The rest of the projected $4.863 million cost was to be "raised from government and special sources." The trustees in February had also approved the purchase of temporary "relocatable" classroom and office buildings to be erected near the gymnasium parking lot.[33]

The "big kickoff" of the sixty-fifth anniversary drive was scheduled for March 19 in Sutton at the Pleasant Valley Country Club, owned by Cosmo E. Mingolla, a college trustee. This event was overshadowed by an important announcement on March 16 from the new Assumptionist provincial superior, Fr. Joseph Loiselle. Speaking on the campus in Greendale, he informed the school's advisory Board of Directors "that the Assumptionists, at a plenary meeting of the North American Province held earlier in the day, [had] voted to close Assumption Preparatory School as of June 1970 and to proceed with the sale of the property."[34] The sale had been quietly but actively explored for several months. The Massachusetts Board of Regional Community Colleges would buy the campus and its buildings for the use of Quinsigamond Community College.[35] The Assumptionists' decision was based on financial difficulties (including a deficit that had reached — with debt service — $400,000 per year), the "scarcity" of religious personnel, and declining enrollment.[36] Because of "pressures applied by the local newspapers which threatened to print rumors,"[37] the decision to close the preparatory school, announced to the Board of Directors on Sunday evening, March 15, was conveyed to the public on Monday, March 16, three days before the "kickoff" of the college's campaign.[38]

There was immediate public speculation that the prep school had been sacrificed for the benefit of the college, with the proceeds of the sale destined for Salisbury Street. Fr. Georges Bissonnette, in fact, publicly expressed the hope that at least some of the $3–4 million the province stood to gain would be shared with the college.[39]

The province's own treasury had been "overburdened" for some time before the decision to close the school. Part of that burden was due to obligations the province had incurred in trying to help the prep school. It had loaned the school $200,000 and had also endorsed a $1 million loan the school had received from a trust company. The province had also used the property of the prep school to guarantee a substantial bond issue of its own, intended to help several of its houses. Fr. Dionisio Solano, assistant superior general, reported to Rome in July 1970 that "once the debts have been paid, there will be a good surplus (in the order of three-and-a-half million dollars) which the province is waiting for to relieve its situation, which is especially difficult after the loans of recent years. . . . The goods of the 'Prep' obviously represent the capitalization of the salaries of our religious over fifty years."[40]

Fr. Joesph Loiselle, A.A.

In the summer of 1971 Fr. Loiselle would send a letter to alumni of the defunct school stating that the sale price had been $6,825,000. Of that, $3.25 million had gone to pay the outstanding obligations of the school. The remaining $3.575 million had been transferred under a court order to the North American province on February 20, 1971. "Of these assets," Fr. Loiselle wrote, "three million dollars were invested to guarantee a bond issue which the Assumptionists had floated in 1968. This bond issue was used to help a number of houses in the Province; foremost among these was the Prep School, whose pressing needs were the decisive factor in floating the bond issue; and the Prep School property was one of its chief guarantees. With that property sold, it was necessary to insure the guarantee by the above amount. Whatever remains will hopefully be used to create a pension fund and an education fund for the religious of the Province."[41]

The closing of the prep school angered some alumni, who pressed unsuccessfully in court to have the proceeds of the sale go to the college rather than to the Assumptionists. In the opinion of Fr. Louis Dion, who was in Moscow at the time but later served as director of alumni relations, the closing and the transfer of assets upset many other alumni, who directed their anger at the Assumptionists. Meanwhile, world events and campus reactions to them cast another shadow over the sixty-fifth anniversary campaign. President Richard M. Nixon's decision in early May to send American ground forces into Cambodia set off a firestorm of protests on campuses across the nation, and a majority of

Assumption students polled by the Student Government Association voted to join a nationwide student strike. Faced with having the campus closed down for a day or longer, the faculty adopted a policy intended to accommodate members of the student body who felt too swept up in the events of the day to bring the semester to a normal conclusion. Those students could arrange with their instructors to receive an "Incomplete" notation and complete their work for a grade by the end of the first month of classes in the fall; seniors could receive a grade of "Pass" based on work prior to May 4, if the instructor so decided; any student could receive a letter grade based on work completed prior to May 4.[42]

While these developments were complicating the fund drive, some fiscal savings were being achieved through reductions in administration and staff. While most went quietly, the director of public information, who had been told his position was being "abolished," wrote a guest editorial in the student newspaper urging that college authorities "call a week-long conference of everyone involved with Assumption College and get everything out into the open," to be followed by "a series of monthly meetings... so that people can know what is going on.... Trust and confidence are what the community wishes to feel.... Something must be done to quell the growing feeling of distrust and confusion which is spreading beyond the low stone walls of Assumption."[43]

At a momentous trustee meeting in June 1970, Fr. Bissonnette had to report that the drive, intended to be completed that month, had brought in pledges of only $676,869, less than half the goal of the campaign. He and Mr. Mingolla were asked questions "concerning the failure to achieve goals." Both, according to the minutes, "cited the depressed state of the stock market and the economy in general, plus the negative reaction of many potential donors to student demonstrations," adding that the campaign had come "to a virtual standstill since the onset of student demonstrations re U.S. involvement in Cambodia." A lay trustee said he had been "confronted with many negative reactions to phasing out of [the] Prep School," and Mr. Mingolla agreed. The trustees voted to extend the contract of their fund-raising consultants for two months, and then to have a newly hired director of development lead the campaign "until at least December 31st." The idea of hiring a professional development officer was later dropped for lack of funds. Although pledges had reached $766,029 by March 1971, and $767, 834 by October 1 of that year, the campaign had failed by a wide margin to reach its goal.[44]

Emergency Measures

For the same June 5, 1970, meeting, at which the unwelcome news of the failing fund drive was conveyed, Fr. Bissonnette wrote an "interim report" on developments in the twenty months since he had taken office as president. He noted that the college had secured from the federal government a grant of $459,664 toward the proposed classroom and office building, but he devoted a substantial part of his report to the structural issues that had produced so many questions on the campus. In his view, it would take time to complete a transition from a hierarchical system of college governance based on religious obedience to "the give-and-take by which decisions are reached by persons bound only by their devotion to a cause and acting without the restraints imposed by religious vows." He pointed to problems arising in the area of campus planning. The fact that the college architect (Norman D. Nault) was also serving as a trustee and a member of the trustees' buildings and grounds committee was, he believed, "open to the charge of 'conflict of interest.'" Such a charge seemed to be "even better founded" in the case of the campus planner (Fr. Alexis Babineau), who not only sat on the board but was chairman of its buildings and grounds committee. The president suggested "as a way of avoiding needless confrontations with the students and faculty," that both the architect and the planner report to a buildings and grounds committee made up of "competent trustees who are not involved in the actual preparation of the plans for the development of the college."[45]

Affirming the "need there is to recognize the sixty-four years of dedication of the Assumptionist Fathers to the College," he recommended a search for ways "that continuity and recognition could be assured without seeming to give a handle to claims of 'collusion' between the religious community on campus and the Board to the detriment of the lay faculty and of the students." Many of the administration's "thorniest problems with the faculty and the students," he wrote, " have stemmed from their inability to understand the need for Fr. [Ernest] Fortin's and Fr. Babineau's presence on the Board, and more specifically, why Fr. Fortin should be Chairman of the Board" when he was also a faculty member. Whether in response to Fr. Bissonnette or for some other reasons, at the recommendation of the provincial superior, the board on that day elected its first lay chairman, T. Curtis Lloyd.[46]

The board had many other grave matters to attend to. Despite the previous insistence by the board's business and fiscal affairs committee on a balanced budget for 1970–71, the trustees approved a budget showing a $262,883 deficit, presented by administrators as the best that could be done "without adversely affecting the school's operation." It would be the fifth deficit in as many years.[47] One of the trustees remarked, according in the minutes, that "if gifts projected are not forthcoming, we are in an extremely serious position." Turning to the vice president for leadership, the board adopted Dr. Remick's recommendations: (1) that the college indefinitely postpone expansion plans other than those for the religious residence; (2) that the federal grant of nearly a half-million dollars for the classroom and office building be returned with the explanation that the project called for more spending than the college could afford; (3) that the trustees announce a freeze on faculty hiring, promotions, and salaries; (4) that the trustees announce the need to increase the student-faculty ratio through increased enrollment or decreased staffing; (5) that the board "evaluate carefully the present administration in terms of the needs of the institution"; and (6) that the college "undertake an intensive effort to broaden the basis of support... among *all* segments of the community."[48]

Dr. Remick and others believed that the college's efforts to broaden the basis of its support were hampered by an indistinct image of the college in the world beyond the campus. Although not himself a trustee, at the request of the board he led a trustee committee in drafting and publishing in August 1970 a promotional brochure designed to convey the "Purpose, Character and Policy of Assumption College." Issued as a "special report" of the trustees, the publication laid claim to the college's classical and Christian heritage, to freedom of inquiry, to the guidance of the community "by a faculty of teacher-scholars," to a core curriculum based in the liberal arts and sciences, to an ecumenically Christian commitment to "a critical examination of the relation of faith to the problems of human thought and action," and to the liberal arts as a basis for acquiring professional or technical skills. (For the first time in such a declaration, no reference was made to the college's French heritage.) A "Policy Statement" asserted the college's private and Assumptionist identity, its expectation that members would act in ways "consistent with its purpose and character," its respect for free expression, its determination to enforce rules and policies while providing means for constructive criticism, and "the College's firm resolve to defend its integrity and safeguard the rights of every member of the college community." The publication concluded with a list of the decisions made by the board on July 5, including the call for a careful evaluation of the administration.[49]

What impact this publication may have had on the outside world is impossible to determine. On the campus one Assumptionist was struck by the attitude of a "distinguished [lay] faculty member" who complained that "the faculty had not been consulted in formulating the purpose of the college." The priest's response to the layman's concern was another manifestation of the gaps that separated various elements seeking to live and work together in the "new college." The priest who noted the complaint about lack of faculty consultation believed that "newcomers, faculty and students," had developed a mistaken notion that Assumption existed to provide "liberal education" — by which he apparently meant a secular education — rather than a liberal arts education. Meanwhile, he remarked, the lay members of the faculty, "all seem to believe that membership in our community gives them the right to decide what will be the purpose of the community. The problem has never been discussed frankly on campus. I'm not even sure if all the A.A.s [Assumptionists] would insist on the original purpose. . . . Either we sponsor a Catholic liberal arts college, or we don't. We cannot allow a secular education to be given under our auspices."[50]

In September 1970, the Board of Trustees took another significant step both to strengthen the role of Dr. Remick and to respond to demands for student and faculty involvement in policy discussions. It created a college council, chaired by the vice president, which included all major administrators other than the president, plus a representative from the faculty senate, one from the student government association, and one from the Assumptionist community. By January 1971, the trustees were looking at preliminary figures that showed an increase in the anticipated deficit to $400,000 for that academic year despite a $400 tuition increase, and they voted to charge another $200 in 1971–72. Under orders from the board chairman to submit a balanced budget, Dr. Remick observed that program cuts would be required, faculty members would have to be released, the Assumptionists on the campus would have to "be employed to the fullest extent," the departmental structures would have to be reorganized, and the number of part-time teachers would have to be reduced. He also planned to reorganize the administration, eliminating the "coordinate college" status of the women and the need for a special dean. On March 26, he reported finding a helpful spirit among faculty members, and especially the department chairs, as they went through the "agonizing and demanding" job of balancing the 1971–72 budget. He listed two faculty members who would be retired, two first-year faculty members who would be terminated, seven part-time teachers to be terminated, and terminal one-year contracts for eight professors (two of them with tenure). He had also removed the position of associate dean for the coordinate college and had notified two other lay administrators that their positions would probably be filled by Assumptionists.[51]

In the midst of these tensions, the campus atmosphere was considerably lightened by the extraordinary success of the men's basketball team. The program had enjoyed a morale-boosting tradition of excellence for many years, but not until March 1971 could *Assumption* magazine carry a cover sketch of players in action under the headline "The Greyhounds: NUMBER ONE in the Nation." The team had won all twenty-two of its games, including one against Holy Cross College, before succumbing to Providence College late in the season. They had earned the exhilarating reward of being recognized at the end of the regular season as the best Division II team in the country by United Press International. Fr. Bissonnette reported to the board in March, "the success of the basketball team has been an invaluable service, generating a confidence in the students that 'things are great at Assumption.'"[52]

Final Scenes

Less than a month later, on April 20, 1971, Dr. Remick convened a special meeting of the faculty to announce his resignation in order to accept the positions of president and program director of the Chautauqua Institution in New York State. He said the "ambiguity in leadership" at Assumption had produced a situation in which "I cannot fulfill satisfactorily the responsibilities of my own calling." He expressed hope that "the removal of myself at this time will allow the school to find the unity and integrity essential to its survival."[53] The executive committee of the board of trustees on May 6 found itself expressing regret at Dr. Remick's decision and discussing evidence of "a long-lasting dissatisfaction on the part of the trustees" with Fr. Bissonnette as president as well as complaints about the president from students and faculty members.[54] Hearing that, Fr. Bissonnette asked the committee if it wished to have his resignation. He was told by the chairman to do nothing pending the return of the provincial superior, Fr. Joseph Loiselle, from Europe.[55] It was agreed that a committee of three would prepare a report for the June 4 meeting of the board.[56]

At the president's meeting with Fr. Loiselle, Fr. Bissonnette was asked to write up a report for the consideration of the committee of three appointed by the executive committee to consider the presidency. He reported that "the ambiguity of leadership seems to have been with us from the beginning" and traced its origins "to the opinion widely held during the summer of 1968 that the acting president of the College and Chairman of the Board [Fr. Ernest Fortin] wished to have Dr. Remick elected President of the College." He presented an account of difficulties he had encountered from the start of his presidency in October 1968 through the establishment of the college council in 1970, which had removed the major administrators from "any official responsibility to the president for the daily management of the college." By then, Fr. Bissonnette wrote, "the ambiguity of leadership seemed well on its way to clarification to the detriment of the president." He sent the report to all members of the board, along with a letter of resignation, just before the board meeting on June 4, 1972. To his cover letter he added a handwritten note: "I believe Friday will see us taking a giant step toward solving our problems." As the June 4 meeting was drawing to a close, the trustees adopted a resolution thanking Dr. Remick for his service, then went into executive session to hear the report of the committee of three. Following that, the board "voted the acceptance of Father's resignation."[57]

There were a number of other resignations taking place in the college that spring and summer. Five professors resigned from the Department of Psychology and Education, one from Religious Studies, and one from the Institute of Social and Rehabilitation Services.[58] Administrators also were leaving, including the business manager, the registrar, the assistant to the admissions director, and the head of counseling services.[59]

Trust Rediscovered

The trustees established a committee to search for a new president and learned in October 1971 that it had held one meeting during the summer and had contacted the Assumptionist provincial, who had presented three candidates as "qualified and available." In the opinion of the chairman of the board, the search committee's budget "should be limited to something less than $500." The apparent absence of a sense of urgency may have been due to a tangible change in the mood of the institution following the departures from office of Dr. Remick and Fr. Bissonnette, the appointment in July of Fr. Wilfrid J. Dufault to serve as acting president, and Fr. Dufault's appointment of Dr. Raymond J. Marion to serve as academic dean. Fr. Dufault had last held the presidency in 1946–47, before being selected as the first North American provincial and later serving as superior general from 1952 to 1969. Dr. Marion was a 1942 graduate of the college who had been decorated with the Silver Star as a Marine officer in World War II, had been appointed to the faculty in 1946, and had received a doctorate in French history and International Relations from Clark University in 1957. Part of the explanation behind the calming effect of Fr. Dufault's succession may be indicated by a letter he addressed to the trustees on September 10, 1971, reporting "a sense of insecurity among the administrating personnel." He attached the draft of a policy statement that would "offer our administrative personnel the assurance that they are not subject to summary dismissal." On October 1, the board adopted the policy.[60]

Dr. Raymond J. Marion

In October, the acting president also reported that "both in administration and among the faculty there was a considerable feeling of uneasiness and insecurity early this summer.... The uneasiness caused by an exceptional turnover in personnel, coming on top of the restrictions in our budget, led some to put into question the very existence of the school." He pointed to "increased trust in the administration," due in part to his "elucidations given on various and sundry occasions about our financial situation" and to the appointment of Dr. Marion as academic dean. Both helped explain the fact that "since midsummer the atmosphere has been improving steadily." The improvement in morale also had to be attributed in part to the success of the admissions office in attracting the largest freshman class (298 students) in the history of the college, bringing the total undergraduate enrollment to 956, an achievement the trustees hailed with a resolution of commendation to Thomas E. Dunn and his staff. Because of all these developments, Fr. Dufault wrote, as well as "the improvement shown in our finances, and because of the rather remarkable good spirit which everyone seems to note on campus, I believe we are justified in being optimistic, not only for this academic year but for the years to come." In June 1972, the college treasurer reported to the board that "all indications point to a profitable year" and that the projected balanced budget for 1972–73 continued to be "a realistic goal." [61]

Thomas E. Dunn

As the school year drew to a close, nineteen years after the tornado and twenty-six years after Fr. Dufault's first year as president, he sent out one of his periodic reports to the college community. "I rejoice with each one of you," he wrote, "faculty, students, administration and staff as the academic year draws to [a] conclusion and we can look back on the fruits of our labors. We are even more confident that Assumption's future is secure. We have all worked through the difficulties and challenges that faced us this year and have come through with very positive results. Without the courage, cooperation, dedication of each one of us, and the grace of God, it would not have been possible. To one and all I am grateful — Assumption is grateful. God's blessings on you for a summer richly blessed with happiness and sunshine. We look forward to welcoming you back in the fall fully refreshed and with eager anticipation to start another great year with and for Assumption!"[62]

Notes

ABBREVIATIONS

AC — Archives of Assumption College, Worcester, MA

AG — Archives of the Assumptionist General House, Rome

AP — Archives of the Assumptionist Provincial House, Brighton. MA

PA — Archives of the Assumptionist Province of Paris

SP — Archives of the Diocese of Springfield, MA

1. Louis Dion, videotaped interview, May 6, 1997, AC; Desautels," Rapport semestriel," Feb. 18, 1954, AP; *Assumption* 25, 4 (Fall 1964): 4–5.
2. *Assumption* 26, 4 (Fall 1965): 10.
3. *Assumption* 26, 4 (Fall 1965): 10; college catalog, 1965–66, 9–10.
4. *Assumption* 26, 4 (Fall 1965): 10; college catalog, 1965–66, 9–10; For examples, see Fortin to Desautels, Sept. 27, 1965; Desautels to Fortin, March 6, 1966; Desautels to Fortin and Alexis Babineau, March 20, 1967, AP.
5. Louis Dion, videotaped interview, May 6, 1997, AC; *Assumption* 26, 2 (Spring 1965): 3; Fortin to Desautels, Sept. 29, 1965, AP.
6. *Assumption* 26, 3 (Summer 1965): 3.
7. House Council Minutes, 58, 68, AP.
8. *Assumption* 27, 3 (Summer 1966): 6–7.
9. *L'Assomption,* n.d. [1954]: 60; Desautels, "Rapport annuel," July(?) 1963, AG 2PN-140; *Assumption* 28, 3 (Summer 1967): 12; *Assumption* 26, 3 (Summer 1965): 12–13.
10. *Assumption* 26, 4 (Fall 1965): 13; *Assumption* 27, 1 (Winter 1966): 23; Leo N. Turcotte, "An Undergraduate's Comments," *Assumption* 27, 2 (Spring 1966): 36.
11. College catalog, 1966–67, 21; college catalog, 1966–67, revised edition, 7–9.
12. *Assumption* 28, 4 (Fall 1967): [1].
13. *Assumption* 28, 4 (Fall 1967): 9–13.
14. *Assumption* 28, 4 (Fall 1967): 22–23.
15. *Assumption* 24, 3 (Summer 1964): 7; *Assumption* 27, 1 (Winter 1966): 1; Alexis Babineau et al., Report to Provincial Chapter, Nov., n.d., 1968, AP; Moquin and Richards, *Assumptionists in the United States* (Worcester, 1964), 92.
16. Louis Dion, videotaped interview, May 6, 1997, AC; Trustee Minutes, April 19, 1968, AC.

17. Trustee Minutes, April 19, June 7, Oct. 11, 1968, AC.

18. June 16, 1968, Superiors folders, AP; Desautels, July 2, 1968, Superiors folders, AP.

19. Trustee Minutes, Oct. 11, 1968, AC.

20. For example, *Assumption* 29, 3 (Fall 1968): 1; *Worcester Telegram,* Oct. 12, 1968, 1.

21. *Assumption* 30, 1 (Spring 1969): 12.

22. Trustee Minutes, April 19, 1968, AC; Lola E. Boyd to the College Council, Jan. 21, 1971, filed with Trustee Minutes, March 26, 1971, AC; Remick, Report of the Vice President, March 26, 1971, filed with Trustee Minutes, March 26, 1971, AC.

23. *Assumption* 30, 1 (Spring, 1969): 2–3.

24. *Assumption* 30, 1 (Spring 1969): 4–7; *Assumption* 30, 2 (Summer 1969): 1; *Assumption* 31, 5 (Aug. 1970): 12–13.

25. Desautels to E. Fortin, Feb. 21, 1967, AP; "Summary of the Discussion on the Faculty residence at the April 7th meeting of the Religious Community," April 6, 1967, AP; "Meeting of Religious Community: Subject: Residence of Religious," April 14, 1967, AP.

26. *Assumption* 28, 2 (Spring 1967): 12–13.

27. "The Meeting of the Religious Community: Subject: The Residence of the Religious," April 21, 1967, AP.

28. Sept. 13, 1970, Superiors folders, AP.

29. Augustin Danby to Paul Charpentier, Oct. 7, 1970, AG JD-43.

30. "Ryan-Letendre Issue 5/69," folder, AC; Bissonnette, Report, May 16, 1971, Trustee records, AC.

31. Bissonnette, Report, May 16, 1971, Trustee records, AC; Memorandum to Faculty, Students, and Administration, Dec. 3, 1969, Trustee files, Bissonnette, AC.

32. "Black Studies Issue — 2/70," folder, AC.

33. Trustee Minutes, Feb. 6, 1970, AC; *Assumption* 31, 3 (May 1970): 3.

34. Minutes, Board of Directors of Assumption Preparatory School, Inc., March 14, 1970, AP.

35. Minutes, Board of Directors, Dec. 14, 1969, AP.

36. Joseph Loiselle to Board of Directors, Feb. 22, 1970, AP.

37. Minutes, Board of Directors, March 14, 1970, AP.

38. *Assumption* 31, 3 (May 1970): 20.

39. *Assumption* 31, 3 (May 1970): 21; Dionisio Solano to Paul Charpentier, July 16, 1970, AG.

40. Wilfrid Dufault, "A Report on the Status of the Prep School," Dec. 30, 1969, AP; Joseph Loiselle to Board of Directors, Assumption Prep School, Feb. 22, 1970, AP; Solano to Charpentier, July 16, 1970, AG.

41. *Assumption* 32, 3 (Aug. 1971): 8.

42. Solano to Charpentier, July 1, 1970, AG; Dion, videotaped interview, May 6, 1997, AC; Remick to Faculty and Students, May 12, 1970, Academic Affairs files, Administration, AC.

43. *Le Provocateur,* May 11, 1970.

44. Trustee Minutes, June 5, 1970, AC; Trustee Minutes, March 26, 1971, AC.

45. Bissonnette, "Interim Report," June 5, 1970, Trustee folders, Bissonnette, AC. Fr. Babineau was also religious superior.

46. Bissonnette, "Interim Report," June 5, 1970, Trustee folders, Bissonnette, AC; Trustee Minutes, June 5, 1970, AC.

47. Bissonnette, "Campus Report," 2, 2 (Sept. 25, 1970): filed with Trustee Minutes, Sept. 25, 1970, AC.

48. Trustee Minutes, June 5, 1970, AC.

49. "Purpose, Character and Policy of Assumption College," [August 1970], *passim,* AC.

50. Nov. 11, 1970, Superiors folders, AP.

51. Trustee Minutes, Sept. 25, 1970, AC; Trustee Minutes, Jan. 3, 1971, AC; Bissonnette, "Campus Report," 2, 2 (Sept. 25, 1970), filed with Trustee Minutes, Sept. 25, 1970, AC; Trustee Minutes, Jan. 3, 1971, AC; Remick, Report of the Vice President, March 26, 1971, filed with Trustee Minutes, March 26, 1971, AC.

52. *Assumption* 32, 2 (March 1971): 4–5, (My thanks to Steve Morris for some of this information); President's Report, March 22, 1971, filed with Trustee Minutes, March 26, 1971, AC.

53. *Le Provocateur,* April 20, 1971.

54. Executive Committee, Report on Interim Activity — April–May, 1971, filed with Trustee Minutes, June 4, 1971, AC.

55. Bissonnette, Report, May 16, 1971, AC.

56. Executive Committee, Report on Interim Activity — April–May, 1971, filed with Trustee Minutes, June 4, 1971, AC.

57. Bissonnette, Report, May 16, 1971, AC; Bissonnette to individual trustees, June 2, 1971, Trustee folders, Bissonnette, AC; Trustee Minutes, June 4, 1971, AC.

58. Remick, Report of the Vice President, Trustee Minutes, June 4, 1971, AC.
59. Acting President's Report, Oct. 1, 1971, filed with Trustee Minutes, AC.
60. Trustee Minutes, Oct. 1, 1971, AC; Dufault to Board of Trustees, Sept. 10, 1971, AC; Trustee Minutes, Oct. 1, 1971, AC.
61. Acting President's Report, Oct. 1, 1971, Trustee Minutes, AC; Trustee Minutes, Oct. 1, 1971, AC; Acting President's Report, Oct. 1, 1971, Trustee Minutes, AC; Treasurer's Report, June 2, 1972, filed with Trustee Minutes, AC.
62. Dufault to College Community, May 19, 1972, filed with Trustee Minutes, June 2, 1972, AC.

Chapter Six

Laymen in the Presidency: Arguments, Adjustments, Accomplishments (1972–1997)

Author's note: Not mentioned in the previous chapter was an event that bears some significance for this history: In September 1970, the writer became a member of the Assumption College faculty. Although the account from that point forward has been based on documentary evidence and will continue to be, it will also inevitably be affected by his experiences as a participant in the life of the college.

On June 2, 1972, the board of trustees for the first time elected a lay person to serve as president of Assumption College. Dr. Pasquale DiPasquale, Jr., professor and director of graduate studies in English at Illinois State University, was selected from a pool of over ninety applicants, including several Assumptionists. The search committee had selected Fr. Wilfrid Dufault as the best qualified of the Assumptionists and Dr. DiPasquale as the leading lay candidate. After extended discussion, the committee had decided to recommend Dr. DiPasquale for the top position and Fr. Dufault as vice president.

However, both names were placed in nomination for the presidency at the June meeting of the board. The chairman, T. Curtis Lloyd, read a petition in favor of Fr. Dufault, signed by twenty-

President Pasquale DiPasquale, Jr.

seven members of the faculty, urging the board to "continue the atmosphere of trust and professionalism fostered on campus by Father Dufault" by naming him president. Another trustee read a letter from the provincial superior of the Assumptionists, Fr. Joseph G. Loiselle, stating that the provincial council believed that it would be advantageous to hire both men and that Fr. Dufault had "spontaneously expressed his willingness to serve as Assistant to the President or as Vice President." The provincial council was therefore recommending that the trustees elect Dr. DiPasquale as the next president. Fr. Loiselle added that the council believed this appointment "would in no way hinder the Assumptionists' role in the College."

Before calling for a vote, the chairman observed that he had "discussed with the two men the team aspects of the appointment" and had "made it very clear that this Board would look to the President for the decisions." The trustees then proceeded to elect Dr. DiPasquale president and Fr. Dufault vice president. This was followed by a resolution in which the board thanked Fr. Dufault for his service as acting president "during a most critical period of the College's existence."[1]

One of the most significant products of the "atmosphere of trust and professionalism" fostered by Fr. Dufault and by the academic dean, Dr. Raymond J. Marion, had been a document issued in February 1972 by a faculty "Professional Standards Committee." Subsequently revised and approved by the Academic Senate, what became known as the Faculty Policy Document established for the first time official standards and procedures for faculty hiring, promotion, and tenure. In Dr. Marion's words, the college could not "follow any longer the rather generally applied rule that eligibility for promotion is tantamount to promotion." Adopted in September 1972 by the trustees for a one-year trial, the document led to the creation of a new faculty evaluation committee and to what President DiPasquale and Dean Marion considered "the most careful evaluation process ever conducted at Assumption." It resulted in nine promotions from a field of thirteen candidates.[2] In January 1974, the trustees approved the Faculty Policy Document for an indefinite period; with modifications, it is still in use today.

The ability to promote and to grant salary increases was made possible by the trustees' decision, also in September 1972, to lift the freeze on raises adopted in 1970, a decision encouraged by the report that the college had realized an operational surplus of $348,000 during Fr. Dufault's presidential term in 1971–72.[3] The moment apparently seemed opportune to further professionalize the college's operation in a variety of ways. In January 1973, President DiPasquale told the trustees that the college had at long last hired a professional development officer, who would report for duty during the summer of 1973. The college also signed a contract with an outside firm to manage the college's food service.[4] At the request of students, the administration also agreed to hire a director of student activities.[5] In May 1973, the president announced that the administration was adopting a new budgeting process developed by the National Association of College and University Business Officers (NACUBO)."[6] Also in that year, in accordance with Dr. Marion's desire to return to teaching, a search was conducted for a new academic dean, and Dr. Richard A. Oehling was appointed to that office effective in the summer of 1973. Under a number of titles, Dr. Oehling would serve as the college's chief academic officer until 1990.

Dr. Richard A. Oehling

The overall financial condition of the college during Dr. DiPasquale's five-year tenure (1972–77) provided grounds for hope, if not for security. Tuition was held at $2,200 from 1970–71 through 1974–75, rose to $2,400 for 1975–76, and was held at that level for 1976–77.[7] Budgets were balanced each year except for 1976–77, when the president anticipated "a break-even year," but auditors subsequently reported a deficit in operations of $102,597.[8]

The Need to Build

One of the most intractable problems Dr. DiPasquale confronted during his tenure was the lack of adequate indoor space on the campus for residence, for student activities, and for faculty and administrative offices. The most pressing need was for residential space for students. In March 1973, the president conveyed to the trustees his wish to avoid a "bind" that would oblige the college "to force large numbers of students to live off campus."[9] In January 1974, he elaborated:

> *"We have two academic departments, Mathematics and Social Rehabilitation [Institute for Social and Rehabilitation Services], in one old house (Cowan) in the northeast corner of the campus. Another old house (Kilgore) in the same corner of the campus is being used to house fourteen students, and we are leasing another house off campus to house another seven students. There are virtually no recreational or lounge facilities for residents in Alumni, Desautels, and Founders Halls. . . . We have a graduate Dean in an 8x12 office; and the Academic and Graduate Deans, who have been authorized to hire clerical assistance, have not been able to do so for lack of space. We need more faculty office space badly. Our Admissions Officer has to have prospective students and their parents wait for their appointments and interviews in the corridor outside his office in Alumni. We have no gym facilities for women. . . . We have a library that must store some 20,000 volumes so that they are inaccessible to students and faculty. Our classroom utilization is phenomenal. . . and it is only with difficulty that we are able to keep the "old religious dining room" [in Taylor Hall] free for meetings of the Board of Trustees."*[10]

Two months later, in March 1974, the trustees approved the building of two townhouse apartment buildings by the start of classes that September. Aubuchon Hall, with a total of forty-eight beds, was the first construction on the campus since the 1969–70 academic year. Those new rooms did not prevent the college from having to rent off-campus property to house twenty upperclass male students. Even that effort left unhoused students who were "provided for temporarily by means of converting a lounge area into living space and by putting some students in the infirmary." The college was also risking the loss of financial support from the federal government if it continued to fail to meet the athletic and recreational needs of female students. To begin doing that, the president said, it would need to convert the four classrooms in the basement of the gymnasium into locker, shower, and dressing rooms, creating in turn a need to replace the classrooms. The college was so short of faculty office space that, the president reported, "our part-time faculty members are now working out of their automobiles." He called attention again to the library, built for 450 students and with 120 seats serving a student body of 1,400. He noted criticism of the library by the New England Association of Schools and Colleges in 1970 and expressed the "fear that we could be in danger of losing our accreditation" if the problem were not resolved by their next visit in 1979–80.[11]

The trustees responded in September 1974 by authorizing $15,000 for a preliminary study of a proposed classroom and office building. Based on the study, in January 1975 the trustee Buildings and Grounds Committee proposed that "a cluster of small buildings" (eventually named Fuller Hall, the Switzer Building, and the DiPasquale Media Center) be built near the Kennedy Science Hall. The board approved $45,000 for those plans and specifications, and also for "alterations of the Gym necessary to meet the requirements of Title IX."[12]

The need for proper student living space involved more than adding buildings. In December 1974, the Assumptionist provincial superior, Fr. Joseph G. Loiselle, writing as chair of the trustees'

Student Life Committee, had reported that "the physical condition of the three large and older dormitories [Alumni, Desautels, and Founders] has deteriorated appallingly and rapidly," due in part to student abuse, but more to the "cutbacks [that] were imposed in the area of maintenance" during the budget crisis of 1970–71 and never restored. The business manager replied that he was including $100,000 in the 1975–76 budget for painting, replacement of tiles, and replacement of some furniture in the old dormitories, and that he would soon be asking the trustees to authorize a room damage deposit from each student.[13]

At the January 1975 meeting, at which plans for the "cluster of small buildings" were authorized, the Buildings and Grounds Committee also recommended the construction of another set of townhouse apartment buildings for an additional forty-eight students (Bissonnette Hall). Although that project was completed by the opening of school in September, the college was still housing forty-three students in six leased houses off the campus, In addition, because of the difficulty of finding more off-campus space, it "had to inconvenience 27 incoming freshmen for the first week or so by putting them in temporary spaces in various locations on campus." By February 1976, the dean of students was projecting a September shortfall of sixty-five on-campus residential spaces, with a maximum of fifty off-campus rentals available. In May, Dr. DiPasquale reported to the board that "we are still in a difficult bind for living facilities." Qualified students were being turned away from the next freshman class because of the need to limit the number of residents, and administrators were facing the challenge of persuading sixty upperclass males to live in leased off-campus residences.[14]

As part of the new planning and budgeting system, the president submitted to the trustees in September 1974 a five-year plan intended, with updates, to guide the college between 1975 and 1980. He described it as "a compilation of the five-year projections submitted to the Planning Committee of the College by various Cost Centers," plus a final page, prepared by the business manager, projecting income and expenses and indicating "ways we can financially achieve what has been set forth as our goals and objectives for the next five years." At that same meeting, the trustees, following an evaluation, "renominated" Dr. DiPasquale for a five-year term, to begin July 1, 1975.[15] By May 1976, the president could point out that the total enrollment was 1,192 full-time undergraduates, compared to 1,197 projected in the five-year plan. Income figures were also meeting projections, thanks to a tuition increase, fund-raising, and the collective financial contribution of $100,000 per year from the Assumptionists — who, if employed by the college, were being paid. The new apartment buildings were, thanks to the room rental fees, self-amortizing, and the cluster of three new buildings, financed in large part by a bank loan, involved a commitment of nearly $1 million, with plans to pay off the loan in ten years.

"Obviously," Dr. DiPasquale wrote to the trustees, "we have altered our attitudes considerably from what they were in July of 1972, when I began my tenure as president. At that time, we were still concerned about the incoming freshman class, we were struggling to put together a faculty policy that would allow us to unfreeze promotions and tenure cases, and there was widespread belief that we had no chance for further growth, since we were in the fourth and last year of transition to a coeducational college. There was certainly no thought that we would be doing any building; there was fear that increased tuition would hurt our enrollments; and there was some concern about whether the federal government was going to continue student financial aid on the same scale that we had been experiencing." The time had come, the president wrote, to leave such apprehensions behind and look to the future, beginning with setting priorities among all the hopes compiled department by department in the five-year plan.[16]

No Capital Campaign

Assumption College did not undertake a capital fund drive during Dr. DiPasquale's tenure, though the need was great and the trustees thought about the possibility frequently. In September 1973, two former presidents were assigned to the development office, Fr. Louis Dion, to work on planned giving, and Fr. Georges Bissonnette, to engage in annual corporate solicitation. One year later, the chair of the trustees' Development Committee alerted the board that "since it is obvious that there are serious space needs, plans must be made for a Capital Campaign, and the Development Committee and the Development Office will recommend such a campaign in the near future." President DiPasquale's five-year plan called for three consecutive fund-raising programs designed to bring in around $17 million. First would come a campaign with a $3 million goal, to help finance a classroom-office complex, a library, and renovation of the Maison Française as an administrative office building. Phase two, planned to coincide with the college's seventy-fifth anniversary in 1979, would have a $4 million goal and would place its emphasis on student facilities. Phase three, according to Development Director Edward A. Engel, would have a goal of $10 million in endowment "and should be considered a never-ending campaign."[17]

In March 1975, the trustee committee announced to the board that it was formally proposing the $3 million campaign, with a "kickoff date" around March 1976. When the board officially approved the campaign in December 1975, the renovation of the Maison Française had been removed from the package; $900,000 was to be sought for the classroom-and-office building and $2.1 million for a new library. The advance gifts and foundations phase was to begin early in 1976, general solicitation in the summer, and the public "kickoff" was scheduled for September. However, in February 1976, the Development Committee raised concerns about competing with a high-profile community campaign to restore Worcester's historic Mechanics Hall. It was decided to continue organizing the campaign, but to put aside any public announcement during 1976.[18]

The whole matter returned to the board in December 1976, when it considered and approved a substitute five-year, $4 million campaign, with $1 million directed toward the "classroom-office complex," $2.5 million toward the library, $250,000 toward gymnasium facilities for women, and $250,000 for renovations to the Maison Française. The advance phase was to begin in the first quarter of 1977, general solicitation in the summer, and the "kick-off" would come in September. However, in May 1977, the president observed that the board was "still being cautious about the kickoff date and the goal to be set for the campaign." By September, when it was reported that the campaign had been further delayed, President DiPasquale had resigned and moved on to a new position.[19]

Structural Tensions: Board and President

During his tenure Dr. DiPasquale also encountered some of the confusion about roles and authority that had created difficulties for Frs. Dion and Bissonnette. On January 5, 1973, the administration asked that the board explicitly delegate to the administration the authority to determine the hours at which male students could visit in the rooms of females and vice versa. Rather than do that, the board noted that the administration had in the past exercised such authority, and "it was agreed" that the trustees' committee on student life would meet with the administration's committee "to follow through on the setting of guidelines." In May, the president reported that the joint meeting of the student life committees had recommended, and he had decided to approve, a change from twelve-hour to twenty-four-hour "parietals" beginning the following September. For students who preferred to live a more restrictive environment, sections of dormitories would be set aside. As a trustee member of the joint meeting, Fr. Joseph G. Loiselle, provincial superior, had submitted to the board a minority report questioning some of the rationale advanced for the policy. He also expressed concern that unrestricted parietals would provide "occasions for promiscuity," that peer pressure might lead students who really preferred restricted hours to opt for the open areas, and that overnight intervisitation would not promote academic excellence. Consequently, he believed that the policy was contrary to the goals and purposes of the college. What the minutes describe as a "lengthy discussion" was apparently brought to a close only when the chairman of the board insisted — apparently mistakenly — that the trustees at a previous meeting had given the administration "full power" to establish the policy.[20]

The issue returned in altered form two years later (in 1975), when students asked the trustees' Student Life Committee to support the elimination of the remaining restricted floors. Fr. Robert J. Fortin, acting chair of the committee, opposed dropping the restricted areas, a judgment in which he was joined by Dr. DiPasquale and the rest of the board, and that settled the matter for the time being. However, in preparing his statement for the board, Fr. Fortin had revisited the 1973 discussions of parietals. He insisted that parietal hours were indeed a matter of college policy and thus should be handled by the board itself rather than the administration, but he found that some trustees in 1973 had in confusion entrusted that responsibility to the administration. He recalled that the discussion on May 26, 1973, had been both lengthy and "heated" because there were "misunderstandings" among board members about the policy-level status of the decision. Implicit in his argument was the proposition that the trustees on that day should have asserted their policy-making authority and either rejected or approved the twenty-four-hour parietals.[21]

There was at least one other structural issue on which Dr. DiPasquale held a view not shared by a majority of the board. In December 1975, he opposed the nomination to the board of an Assumptionist priest who was also a member of the faculty. He argued that the presence of a faculty member on the board would, in the words of the minutes, "open the door to lay faculty members and then students being nominated for Trusteeship." The Bylaws and Nominating Committee had not discussed this issue, said its chairman, because "there has never been a policy relative to a candidate's position on the faculty." Some trustees argued that "ample time" should be afforded to address this new policy question, while others argued that the candidate's nomination was based not on his faculty status but on his membership in the Assumptionist community. In the end the candidate was elected to the board, with three trustees voting against it and another abstaining.[22]

Structural Tensions: Board and Faculty

A more extended debate over the domain of the trustees began to emerge in May of 1974. The faculty had spent many hours during that academic year discussing alternatives to the distribution system adopted in 1968 as the core curriculum, but reached no consensus. The chairman of the trustees' Academic Affairs Committee, who served as a professor at another institution, told the board "that he considered it most important that the Faculty not alter its distribution requirements without first inquiring into the meaning of liberal education per se, and the meaning of liberal education at Assumption College." In his written report, he argued "that the study of the great writings of the Graeco-Roman and Christian traditions, and the raising of questions that will help us act in the world as men [*sic*] and citizens must somehow provide the basis or framework of education at Assumption."[23] Although no formal action was taken on his proposal, this approach to curriculum would play an important part in subsequent deliberations.

On March 21, 1975, the trustees received the Academic Senate's proposal for a new core curriculum. Under "Skills" it called for demonstrating ability in English composition, in mathematics, and in a foreign language, at levels to be determined by the faculty. Under "Philosophy and Religious Studies," it called for two unspecified courses in philosophy or religious studies or one in each. Under "Humanities, Natural Sciences, and Social Sciences," it required three unspecified courses in different areas of the humanities, two in natural sciences and mathematics from any of five listed areas, and one course in each of three social sciences. The trustees' Academic Affairs Committee recommended that the board require three courses rather than two in philosophy and religious studies, four courses rather than three in the humanities, that Computer Science be dropped from the list of specified science and mathematics courses, and that geography be stricken from the social science list. The chair of the Academic Affairs Committee thought the trustees should also go further and make a statement about the content of the required courses.

According to the minutes, President DiPasquale expressed strong approval of the faculty proposal, though he preferred three rather than two courses in philosophy and religious studies. He objected to eliminating computer science and geography and said he did not think the board should comment on the content of courses. Another trustee agreed with the addition of a course in philosophy and religious studies but questioned the board's right to eliminate computer science or geography. Another "felt that the Board of Trustees should confine itself to matters of policy even where curriculum is concerned and should not go into details of curriculum."[24] A trustee later observed that, at this meeting "several members of the Board [had] objected to the form in which votes were taken because they combined both generalized policy with which the majority agreed, and very specific objectives, which a minority of the members felt were minor issues that did not rise to the stature of policy and, therefore, should not be voted by the Board."[25]

When the votes were taken, all of the changes recommended by the Academic Affairs Committee were adopted by the trustees. The board also adopted three statements offered by the committee:

1. *Courses should draw mainly from the disciplines some of their essential contributions to the understanding of the world, or to the guidance of human aspiration, conduct and taste.*
2. *Courses (all courses) should employ, wherever feasible, established classics of Philosophy, Religion, Literature, Art, History, Science, etc., and encourage their careful study and appreciation.*
3. *Courses should be developed for the well-educated man [sic] or citizen, and not simply for the major and the discipline as such, and should not mainly stress the methods or techniques of the disciplines.*[26]

A few weeks later. the Academic Senate agreed to the additional course in philosophy or religious studies, recognizing "the special concern of the Board... with insuring 'due prominence' for these two disciplines in the curriculum." However, the senate defended its position on the other courses, adding that "those decisions should be accorded a respect commensurate with the professional expertise and the arduous process which created them" and urging the board "to continue to entrust to the judgment and competence of its faculty the determination of specific details of the curriculum." On May 16, upon the recommendation of its Academic Affairs Committee, the board restored geography to the list of social science options but left the other changes in place as matters of college policy.[27] The new requirements went into effect for the 1975–76 academic year.

Business Studies and the Liberal Arts

A matter of considerable importance for the college's future gained some prominence in August 1976 when the trustees attended a "seminar weekend" on the campus to explore with major administrators a wide array of issues. One topic was the perceived need to inaugurate new programs to attract students. Considerable attention was given to the potential attractiveness of "programs in business and management." The college's concentration in business administration, which had been instituted by Fr. Desautels with considerable publicity in 1953, had remained in the catalog until 1960–61, when it had been replaced by a concentration in economics. The seminar participants noted the popularity of the major in Social and Rehabilitation Services and turned to a broader examination of "job-oriented programs in general and whether or not they were appropriate to a liberal arts college." At the September meeting of the board, Dr. DiPasquale reported that the English Department had responded with "enthusiasm" to the idea of adding more career-related programs and that he had asked the Economics Department "to seriously consider a Business Administration program." Enthusiasm was evidently not universal among trustees, the faculty, or the Assumptionists. In his report to the board in May 1977, the president took some note of the opposition this idea had generated. "I am not suggesting that we should initiate fly-by-night programs to answer the passing fancies of students just to get them to enroll here," he wrote. "Quality should be our first concern in planning a new academic program.... I am also not suggesting that we should in any way sacrifice the liberal arts character of the institution. If instituting a new career-oriented program in any way disturbs members of the faculty, the Board, or the Assumptionists, then I suggest that thought be given to improving the liberal arts component of our undergraduate education."[28]

Three months earlier, on February 25, 1977, Dr. DiPasquale had submitted his resignation from Assumption in order to accept the presidency of Loras College, a diocesan institution in Dubuque, Iowa, where he would become the first lay president. The chairman of the board, William E. Aubuchon, Jr., appointed Fr. Wilfrid Dufault to serve once again as acting president until a successor to Dr. DiPasquale could be appointed. He also instructed the members of the Bylaws and Nominating Committee to act as a search committee and recommended that they set up a "Committee B" made up of two trustees, two administrators, two faculty members, two students, and two alumni to carry out various duties, including initial screening of candidates and presenting "several candidates" to the Bylaws and Nominating Committee, which was, in turn, to submit no fewer than two names to the board.[29]

Among the duties of Committee B was the drafting of a "presidential profile." The draft submitted on April 28 opened with a preamble that focused on the "series of challenges" the college was facing, including "prospects of diminishing enrollments in the early 1980s, the ongoing need for curricular development in new academic programs and in the core program of liberal studies, the need for further development of campus facilities, and the related need for sound fiscal management and energetic fund-raising." It divided the qualifications into two categories. The first called for evidence of high academic achievement and the hope that the candidate might "be born out of the liberal arts tradition." The second called for a person who would "understand and support Catholic higher education," which included "familiarity with the history of Catholic education, an understanding of what makes a school Catholic, a certain knowledge of Catholic doctrine, and a willingness to apply these principles in the day-to-day leadership required of a college president."[30]

Dr. DiPasquale in his final report to the board reacted strongly against the profile's forecast of diminishing enrollments. "I still cry out loudly that that is the wrong approach and the wrong way of thinking about the future," he wrote. Although across the nation the number of eighteen-to-twenty-two-year-olds was "going to level off and then drop slightly," he contended that "Assumption's enrollments will be what the college wants them to be, as long as the Admissions Staff remains as competent as it is, as long as planning is careful, and as long as we have the programs and campus life that students want."[31] Committee B's draft also attracted a complaint from Fr. Edgar Bourque, the provincial

Fr. Edgar A. Bourque, A.A.

superior since Fr. Joseph Loiselle's May 1975 appointment as assistant superior general and treasurer of the worldwide Assumptionists. Fr. Bourque observed that if being a Catholic was not a requirement for the presidency, that "could make a difference to those who contribute their services." The profile was promptly amended to require that the new president be a Catholic.[32]

Fr. Dufault Returns

As the search for a new president got under way, Fr. Wilfrid Dufault took up the duties of the presidency for the third time in his career. His September 1977 report to the board of trustees was firm, yet somber. The major administrators, he said, were working in a spirit that inspired "confidence that we will overcome our shortcomings and face present challenges with courage." He recounted efforts to find housing for students after it became clear that an unexpectedly large freshman class of 409 was expected. He was also struggling with the consequences of the budget deficit experienced during the final year of Dr. DiPasquale's tenure. The problem was blamed principally on a decline in enrollments in the graduate school, following changes in state certification requirements for teachers. The budget shortfall had resulted in another freeze on salaries and hiring, and the acting president declared that he saw "salary raises as our very first priority."[33]

Fr. Dufault noted that the long-delayed $4 million capital fund drive would not become public in the near future, because the operating deficit would drive away some potential donors, and the campaign had failed to attract a single major gift. To increase income he urged continuous attention to the curriculum to "make sure that we continue to deserve the good reputation Assumption enjoys. This means striving to improve both our liberal arts core and our peripheral offerings. Then we must strive to appeal to a broader spectrum of college-age youth by introducing a new peripheral program in great demand. Such is the case of the business program." He urged similar attention to the market for graduate and summer school courses, but put special emphasis on "rebuilding" undergraduate education for adults, as had been intended following the phasing out of the St. Augustine Institute. He promised a study of "the demand for adult education in our area and how we could meet it."[34] The interest in continuing education stemmed in part from the demographic studies forecasting that the number of Americans between eighteen and twenty-two years old would shrink between the early 1980s and the mid-1990s, reaching a 15 percent decline nationally and 30 percent in the Northeast by 1995.

In December 1977, the trustees, ever aware of the acute need for new facilities, authorized a $7.5 million campaign to fund "the existing classroom-office building complex [Fuller Hall, Switzer Building, the media center], extension of the gymnasium for women's facilities, student center, library, additional classrooms and offices, outdoor athletic facilities, and renovation of the Maison." No public announcement was to be made without the express approval of the board.[35]

President Hagan Introduces Himself

The financial problems, the capital campaign, the dearth of physical facilities, the projected enrollment declines, the promise of continuing education, a fresh examination of the core curriculum, the introduction of a major in business studies, all were among the challenges awaiting the next president. On May 19, 1978, the trustees conferred that office on Joseph H. Hagan, special assistant to the chairman of the National Endowment for the Humanities in Washington, D.C., and former vice president of Bryant College in Smithfield, Rhode Island. Addressing an opening convocation at the Mass of the Holy Spirit in

President Joseph H. Hagan

September, the forty-three-year-old president emphasized the Catholic character of the college, echoing many speakers in the history of the college in declaring that, in addition to studies and careers to be thinking about, "we have an eternal destiny with God, which is the overriding purpose of our existence. If we should gain all else and forsake this, then no matter how successful our lives may be by contemporary standards, we have failed completely. This may not be a popular concept today, but it is essential to an understanding of the Catholic faith and the Catholic college." The president complimented the faculty, staff, administration, and student body before turning to consideration of the college's problems.

For Assumption, and for most other small liberal arts colleges across the country, he said, "the single biggest problem is money.... This school is in no immediate danger of going out of business, but our continuance demands a tight, lean operation, and the good will of everyone. There is no other way."

The new president closed with a declaration of firm support for the Assumptionists and their roles in the college and his hope that more of them would join the faculty, administration, and staff in the years ahead.[36]

The Money Problem

President Hagan attacked the money problem from the start. He decided early to reorganize and strengthen the business office, combining the positions of treasurer and business manager into one and adding a comptroller and a personnel manager. He also launched an extended effort to put the college on a course that would enable it to carry out realistic and successful capital fund campaigns. He began that long process by insisting that the $7.5 million capital fund drive be delayed until a "national fund raising consulting firm" could carry out a feasibility study. In this context, the new president also objected to a provision in the bylaws that excluded him from serving on the trustees' Bylaws and

Nominating Committee. He argued that the president should play a major role in the selection of trustees. As a compromise, he settled for language stating that the president would be invited to committee meetings and would have a "key role to play" in the selection of trustees. He also let the trustees know that during the feasibility study for the capital campaign "each trustee will be visited by a senior officer of the firm selected in order to determine his interest in the campaign, his willingness to contribute to the extent of his financial ability, and his readiness to solicit others."[37]

In March of 1979, the president told the board that during the previous fiscal year, although the college had collected $338,000 in outside funds, "only $2,700... was contributed by the Board. This is less than 1% of the total, and even the most conservative fund raising counsel estimate that trustees should account for between 20-30% of a capital fund. There is something seriously wrong when this is not the case." The consultants reported in May 1979 that a $2 million drive seemed feasible, and the board approved further preliminary steps to prepare for a vote in December. In October, President Hagan told the board that so far five of them had made pledges "in the five figure category" and that "your full support is both needed and appreciated." In December, the chair of the Development Committee observed that the response of some trustees had been disappointing, but the board agreed to let the silent phase continue and set the goal as $2 million to support much-needed dormitories and a campus center. In March 1980, President Hagan again asked trustees to finalize their pledges, reminding them that "100% trustee participation at responsible levels is very important to these foundations."[38]

More than a year later, after authorizing a bank loan for the construction of student housing, the trustees' Executive Committee called for the capital drive to continue with a campus center as its principal goal. In October 1981, the president reported that $900,000 had been raised in cash and pledges. In March 1983, 100 percent trustee participation had been secured ($406,000 from lay trustees and $100,000 from the Assumptionists). With $1.6 million raised in cash and pledges, the board authorized the launching of a public campaign for $2.5 million, $1,800,000 to support construction of a campus center, $200,000 as an endowment for the management of the center, and $500,000 to meet a challenge grant from the National Endowment for the Humanities to support salaries and faculty development. At the same meeting, the trustees authorized the start of construction of the campus center that spring.[39]

In his March 1983 report, the president reminded the board that when he came to Assumption in 1978, "there were tentative plans to begin a public capital fund drive for $7.5 million," and he had been told that the campaign for a student center was to be announced that September. "As most of you know," he wrote, "I applied the brakes to both projects." Now that success was at hand, the college could build upon it. "Nothing succeeds like success, and if Assumption is to have ultimate success in its current and future fund-raising activities, we must convince people that we can raise large sums of money and that we will expend those monies on the projects intended by the donors." The campus center — later named for President Hagan and his wife, Patrice O'Malley Hagan — opened at the start of the spring 1984 semester. By February 1985, the campaign goal of $2.5 million had been achieved, with a $9,000 surplus. The trustees had contributed $600,000 of the total.[40]

The lessons learned between 1978 and 1985 were soon put to good use. As the campus center campaign concluded in February 1985, the board was authorizing a feasibility study for a campaign to build a new library. In October, it approved what turned out to be a $3.5 million effort. The Emmanuel d'Alzon Library opened in the fall of 1988. By early the following year, the campaign had raised $3,543,595 and funds were still being pursued to support an operating endowment for the library.[41] The library campaign had brought in over $3.8 million by the time a new feasibility study led, in early 1990, to the start of a three-year, $15 million campaign for a student recreation complex, renovations to existing buildings, endowment, and supplementing the annual fund over three years.[42] Construction of the $5.8 million Plourde Recreation Center (named for trustee Gloria Plourde and her husband, Roger) began in September 1991, and the facility was opened in January 1993.[43] By the spring of 1996 the $15 million campaign goal had been surpassed.[44] The largest gift was from the Assumptionists, who donated $1.375 million to establish and endow the d'Alzon Chair in Liberal Education.[45]

Financial Improvement

Francis P. Gurley

While they were working on the capital campaign agenda, President Hagan and his administration — in particular Treasurer Francis P. Gurley, who joined the administration in 1979 — also had to confront "money problems" in the daily operation of the college, problems similar to those faced by President DiPasquale. The basic challenge was to grow slowly while maintaining a balance between the need to spend money on overdue projects and the need to keep tuition costs within the financial capacities of the students and their families. Less urgent, but still important, was the need to overcome the deficits accumulated over previous years of financial struggle and to build an endowment that might someday allow the college to be less dependent on tuition and fees.

After suffering a deficit in the 1976–77 academic year, the trustees had refused to consider raises for employees the next year. Partly as a consequence, the college realized a $111,000 operating surplus during Fr. Dufault's year as acting president. Through tuition and fee increases, outside fundraising, and cautious management, the Hagan administration would continue to produce operating surpluses, rising to $1,741,810 in 1989–90, and remaining over $500,000 through the mid-1990s. Like Fr. Dufault, the new president urged attention to the financial needs of employees, especially the faculty. He hoped to provide remuneration within the middle range of national surveys of faculty salaries, noting in 1978 that the college was in "the lowest rating on the salaries of assistant professors, associate professors, and instructors." By the time he wrote his five-year report in 1983, President Hagan could claim to have "increased faculty, administrative and staff salaries to what is now a very respectable level."[46]

In other ways as well, the financial position of the college improved markedly during President Hagan's tenure. Between 1973 and 1978, the college had withdrawn nearly $500,000 from its endowment fund to meet current needs.[47] As of February 1979, the fund was reduced to just under $520,000.[48] In October 1984, as a result of the annual surpluses, President Hagan was able to announce that the accumulated operating deficit had been reduced to zero.[49] In 1985, the president asked the trustees to note that "Worcester area banks, once very hesitant to loan us money, now compete for our business."[50] Even so, in October 1986, the president could announce that the college had gotten through the 1985–86 academic year without once having to resort to short-term borrowing to meet obligations, a dramatic triumph over the cash-flow problems that had dogged the DiPasquale administration.[51] In February 1988, he noted that the budget contained provisions for reserve accounts. "In a sense," he remarked, "we are now going beyond the point of 'hand to mouth' funding and are strengthening ourselves in the process."[52] By 1989, the net operating surplus had risen to $6 million, a major reason for the increase in the endowment to $6,060,980.[53] Five years later, the total had doubled, and by the time President Hagan left Assumption at the start of 1998, the endowment fund held more than $25 million.

Housing and Student Life

President Joseph H. Hagan and Mrs. Patrice O'Malley Hagan

From their earliest days at Assumption, President Hagan and his wife, Patrice O'Malley Hagan, took a strong interest in the lives of students. The new president began immediately to schedule a weekly breakfast meeting with student leaders. With his wife he visited student residences for receptions and dinners, "generally making ourselves visible and accessible to the students." When the couple moved into the newly purchased President's House in 1979, they began holding evening receptions to meet and talk with groups of around 100 students at a time. "It will take several months to visit with all 1400 students," the president wrote, "but we feel it will be an extremely effective vehicle for improving communication and strengthening student morale." He strongly supported the efforts of a new dean of student affairs, Fr. Peter Precourt, to bring "a new sense of order, enthusiasm, and purpose to the Student Affairs area of the college." He added to the staff in the student life office and pressed for members to be present on the campus during the evening hours. When the trustees' Bylaws and Nominating Committee evaluated the president after his first two years, it observed, "The students are very important to the President. From the outset, he was concerned about improving the quality of their life. There is great satisfaction with the changes he has made in the office of Student Affairs and with the direction that office has taken."[54]

In the president's dealings with students, and in the year-to-year management of the college's finances, student housing continued to be a central concern. The residential situation, President Hagan said in his first report, "continues to be extremely crowded," and the conditions "adversely affect students' academic performance and adjustment to college." The Executive Committee of the trustees initially insisted that funds for housing should be included in whatever capital fund drive might emerge from the feasibility study the new president had set in motion. In March 1979, however, the board approved borrowing for student housing, provided that no operating funds be used and that financing "not be a drain on the College." During the first week of the spring semester in 1980, new apartments (Moquin Hall and Dion Hall) were occupied by 108 students. That addressed some of the backlog, but in March the president warned that "some form of overcrowding or new facilities for housing seems imminent for Fall 1980.[55]

The need for housing continued to compete for priority with other facility needs. In early 1980, longtime Athletic Director Andrew Laska, arguing for a playing field to accommodate both field hockey and softball, noted once again that "women have been brought on campus without any major addition to the athletic facilities either indoor or outdoor since 1963."[56] The "third athletic field" (Marois Field) was completed in 1986, thanks largely to trustee Normand J. Marois' donation of materials and labor. Many improvements to athletic facilities would be provided over the years through gifts of money, materials, and labor from supporters of the athletic programs.

Fr. Peter R. Precourt, A.A.

Always conscious of the predicted decline in the number of eighteen-to-twenty-two-year-olds during the 1980s, both the board and the president hesitated to build additional housing even though the student body was expanding (from 1,307 in 1977–78 to 1,464 in 1980–81). In March 1981, the trustees discussed the possibility of building temporary housing that would serve for three years, during which the college could "determine the demographic impact" on enrollment. By a one-vote margin, that course was rejected. At the conclusion of a long discussion, the board approved borrowing for new permanent housing.[57]

Although students were able to occupy two new apartment-style dormitories in September 1981 (Dufault Hall and Authier Hall), there continued to be 105 other students "directly affected by tripling," and the administration was warning of a shortage of 138 beds in fall 1982. For a few years, pressure to build more housing was limited, if not alleviated, by fairly constant enrollment (1,464 in 1980–81, 1,478 in 1981–82, 1,460 in 1982–83). After extensive rehabilitation of the "lower dorms" (Alumni, Desautels, Founders) in 1983, President Hagan said the community could at last "point with pride" to the housing conditions on the campus. Built for 1,077 students, the dormitories were housing 1,095.[58]

This equilibrium was short-lived, as there were again eighty "tripled" rooms on the campus by October 1984. Although consideration of the demographic realities continued to discourage plans for new housing, there was growing concern among trustees, especially Assumptionist trustees, about the academic effects of putting three students in rooms designed for two. That concern was intensified when the president, in February 1985, predicted that, "without additional dormitory space next year, almost all of the freshman class will be tripled all year." The trustees' Student Life Committee, chaired by an Assumptionist, reported that the tripling situation was "getting more and more intolerable." The board authorized borrowing for a new dormitory (Salisbury Hall), which was under construction by May 1985. President Hagan reported that flexibility was being built in by providing double rooms that could "accommodate three easily if that becomes necessary."[59]

Relief was again short-lived. In October 1986, the president reported that Assumption had "unexpectedly welcomed the largest incoming class ever." Just over 500 freshmen had arrived, providing an excess of residents over beds of sixty-five. Students were tripled and quadrupled, and once again lounge and study spaces in Alumni and Desautels Halls were transformed into bedrooms. At that point, the administration and trustees decided to take on the residential needs and the pressing need for faculty office space at the same time. The trustees authorized the expenditure of nearly $3 million for a new 234-bed dormitory (Worcester Hall) to accommodate the extra students, plus those who would otherwise have been destined for Founders Hall. The empty rooms in Founders would be converted to faculty offices and the headquarters of the computer services department. The construction and the conversion were carried out in 1987.[60]

Despite the demographics, enrollment levels at the college began to rise again, to 1,564 in 1985–86, 1,619 in 1986–87, and 1,718 in 1987–88. In early 1988, the board authorized construction of a new 160-student dormitory (Wachusett Hall) that would open in the fall. This led to fresh questions among trustees about whether the new dormitory would "solve" the overcrowding problem; whether temporary overcrowding — temporarily relieved by new construction — should be standard practice as the student body grew. They also asked how large the trustees should allow the college to become, and just how the college did its planning. President Hagan responded to these questions early in 1988 with a number of planning-related initiatives: He recommended creation of a Trustee Ad Hoc Committee on Future Enrollment to consider the question of size, facilities, and finances; he announced the impending appointment of a director of institutional planning; he announced that the administration would begin a formal strategic planning process under the guidance of a consultant; and he announced creation of a special task force of senior administrators to "assure a minimum of overcrowding and a minimum of empty beds" in the fall. Questioned by an Assumptionist about the board's request to keep the next freshman class small enough to avoid overcrowding, Treasurer Francis

P. Gurley said the goal was to enroll only 450, but, if the class came in smaller than that, the college would have to deal with the financial ramifications for the next four years.[61]

A few months later, President Hagan observed that with the freshman class held to 450 students, normal attrition would lead to empty beds — and thus lost income — by January. And indeed, in February 1989, the administration reported that there were seventy-five empty beds, representing a loss of around $300,000 in revenue. The president said his administrative task force had recommended that, in the future, the school year should open with around fifty-seven extra resident students. That, he added, would be the administration's policy unless otherwise directed by the board.[62] The board did not object, and that October the president reported that there had been seventy more students than beds at the opening of classes.[63] The overcrowding problem disappeared for a time in the 1990s as enrollment declined from 1,839 in 1989–90 to 1,649 in 1996–97.

Assumptionist Sponsorship

During the first few years after the creation of the lay-clerical board of trustees in 1968, a variety of questions had arisen about the meaning of the continuing Assumptionist "sponsorship" of the college. In particular, how would the Congregation carry out that sponsorship when the religious occupied only one-third of the seats on the policy-making board of trustees? In January 1972, the trustees and the North American Province of the Assumptionists signed an agreement intended to delineate "some of the mutual responsibilities underlying the notions of auspices and sponsorship." That agreement was superseded by a new one in May 1977, aimed at affirming and strengthening the Assumptionists' "primary role in guaranteeing the Catholic character of Assumption College and in promoting a philosophy of education which is consonant with the highest Christian and classical values and ideals." It provided for Assumptionist representation on search committees for major administrators and for consultation regarding representation on "all committees of the administration." It also provided that the college would "actively seek the services of qualified Assumptionists" and that the provincial superior would encourage as many qualified Assumptionists as possible to work at the college. Among many other provisions, it also called for preferential hiring of Assumptionists and stated that "the Assumptionists shall be particularly responsible for the department of Theology/ Religious Studies."[64]

While this new discussion of the Catholic character of the college was getting under way, the institution took steps to commemorate and reaffirm the other feature that had for so long been a distinguishing mark, its French heritage. In 1979, the trustees approved the establishment of the French Institute of Assumption College. Under the direction of Dr. Claire Quintal, the new institute was to be financially self-supporting as it carried out its mission "to foster the study and appreciation of the people, language, literature, history and culture of France and French North America."[65] On March 15, 1980, the institute sponsored the first in a series of annual conferences exploring the history and culture of the French-speaking peoples in North America. Located at first in La Maison Française, the institute's offices and documentary collections were later transferred to the new Emmanuel d'Alzon Library.

Also in 1979, the long-anticipated Center for Continuing and Professional Education began to function, under the direction of Dr. Richard Greene. Expecting 125 students when it opened its doors in January, the center attracted an initial student body of 180, of whom 65 were candidates for associate's or bachelor's degrees.[66]

In October 1979, President Hagan notified the board that one Assumptionist was serving as acting dean of students, and another as acting assistant dean of academic affairs. The "acting" status was a temporary arrangement, he said, brought on by the objections of the Representative Faculty Senate (which had replaced the faculty-administration Academic Senate in September 1978) to hiring faculty or administrative officers without an open search. The president and the local religious superior, Fr. Robert J. Fortin, had offered, as a compromise, the appointment of a consultation committee to evaluate the qualifications of the Assumptionist nominees. However, the senate had failed to appoint members to a consultation committee.[67]

The president told the trustees he would if necessary make the appointments without faculty approval, but he put the board on notice that "there will continue to be this kind of problem with the faculty over the Agreement insofar as appointments are concerned." In the ensuing board discussion, Fr. Edgar A. Bourque, the Assumptionist provincial superior, warned, according to the minutes, that "if there is too much hassle and the point is reached where Assumptionists don't want to come to the College, it will no longer be an Assumptionist-sponsored institution." Fr. Robert Fortin, the local superior, was recorded as saying "that the faculty has never come to grips with the implication of a Catholic institution, that there has to be a role for the sponsoring body."[68] On October 25, 1979, the council of the Assumptionist province discussed the problem that "the present Agreement does not seem to be practically effective." "In the months ahead we will have to address this problem, which is indeed serious," the council said. "With this Agreement, can we as Assumptionists be faithful to our responsibilities at Assumption College?" The members of the council unanimously passed a "vote of support" that declared, "We recognize that there is a serious problem, and solutions must be pursued energetically, especially now by Fathers Edgar[Bourque] and Robert [Fortin], but eventually by all of us."[69]

On November 2, 1979, the Faculty Senate discussed with President Hagan the trustees' agreement with the Assumptionists and decided to ask for a meeting on November 30 with the chairman of the board and the local superior to discuss the implications of the agreement. The senate repeated its request in February, March, and April 1980, only to be told by its officers that they had been unable to arrange the meeting. At its first meeting of the new

Fr. Robert L. Fortin, A.A.

academic year, on September 12, 1980, the Faculty Senate unanimously reaffirmed "its desire for a meeting concerning the character of Assumptionist involvement in the life of the college." On September 26, the senate president said efforts to schedule the meeting were ongoing.[70]

As early as April 2, 1980, Fr. Robert Fortin had written a substantial, confidential paper on "The Assumptionists and Assumption College." It examined the problems of the 1960s and the thinking associated with the great structural change of 1968, the institution of the lay-clerical board. The Assumptionists, he observed, did not insist on holding a majority of the seats on the board because that "would have been too taxing on our numbers." Nevertheless, he wrote, even in 1968 there were some who thought that, rather than accept a minority of the board seats, the Assumptionists "should have gone the bicameral route in order to maintain adequate control." (In a bicameral system a small upper body controlled by the Assumptionists could have served as a check on the Board of Trustees.) The failure to secure such control had resulted in a situation in which "no one has to take the Congregation seriously because no one is accountable to it. And this is generally very well known on campus!" Fr. Fortin believed that "we — including Assumptionist administrators — have tenured faculty members who are passively and even aggressively inimical to the College's religious commitment. This is serious and will haunt us for years to come. We must find ways of stemming the tide, lest we discover in the not-too-distant future that we are sponsoring a secular college."[71]

Fr. Fortin presented a revised version of this paper in his long-deferred meeting with the Faculty Senate on October 17, 1980. He argued that the Church held the Assumptionists responsible for the Catholic character of the college, and that to carry out that responsibility the Assumptionists "must have at their disposal a way of saying what is or is not consistent with the goals and purposes of a Catholic institution conducted under their auspices. They must have a way of holding people accountable in these matters." In order to exercise this control, he said, "the Order must be able to function corporately within the College structures, and the legislation pertaining to them must be seen as essential, not peripheral, to the institution."[72]

On December 10, Fr. Fortin presented to the Bylaws and Nominating Committee of the trustees (which he chaired) a draft revision of the college bylaws, drawn up by a committee of the Assumptionists. Its most salient feature was a new "board of overseers" that would, among other duties, initiate any changes in the bylaws; have a veto over the election, reelection, or removal of trustees, the president, vice president, and treasurer; have a veto over capital fund drives, disposal and mortgaging of property, mergers and consolidation, and closing of the school; and have the right to ratify or reject any actions of the trustees. The board of regents would be made up of six Assumptionists, and there would be a twenty-four-member board of trustees that would hold "real, though delegated, authority." Included in the twenty-four would be the six Assumptionist overseers, the president, and seventeen non-Assumptionists. By late January, when a draft Assumptionist proposal was released to the campus community, the board of regents had evolved into a ten-member body, evenly divided between Assumptionists and non-Assumptionists. The provincial superior, who would chair the regents, would appoint the Assumptionist members, and the others would be appointed by the Board of Trustees, subject to the provincial superior's consent. The Board of Trustees would include the ten regents and eleven to twenty-one others, possibly including Assumptionists. Other "fine tuning" occurred as deliberations proceeded. Because the proposed structure would give the Assumptionists as a group a controlling role in the governance of the college, the proposal also called for elimination of the agreement between the college and the Assumptionists and the transformation of its provisions into resolutions of the board.[73]

President Hagan appointed Dr. Raymond J. Marion to chair a thirteen-member lay consultation committee consisting of the three officers of the Faculty Senate, two additional faculty members, three senior administrators, three student government officers, and the president of the alumni association. On March 16, 1981, the committee unanimously recommended that the proposed restructuring not be approved by the board. In an attempt to empathize with what they perceived as the Assumptionist point of view, the committee expressed appreciation of the difficulties the religious faced in their role as sponsors and an awareness of "unresolved differences that remain in any community" that had gone through the events from the tornado to the closing of the prep school. The committee also recognized "the many problems which confront an Order seeking to replenish its ranks, maintain its educational mission, while at the same time respecting the vocational choices and commitments of its members" and said it understood "the spirit of renewal which the Order seeks for itself and the important role of the College in that renewal."[74]

"At the same time," the committee said, it recognized the "very substantial progress" the college had made since the creation of the lay-clerical board and "most importantly, [the decision] to rely upon committed and experienced lay men and women in the management of its day-to-day affairs." The committee speculated that the Assumptionists might be reacting to problems occurring at the board level, since "we and our colleagues see no evidence of Assumptionist personnel being treated unfairly or without respect." The committee said "there is no opposition to preferential hiring of Assumptionists within the professional standards maintained by the college" and warned that adoption of the proposed bylaws might undermine "the good rapport which exists between the religious community and the laity of the College." It concluded by informing the board that the proposal had led to "considerable interest, apprehension, anxiety, and some anger among students, faculty and members of the administration.... Many are puzzled about the rush of events and find it difficult to accept the need to arrive so quickly at a resolution of fundamental differences that continue to separate the perceptions of the proposers of the changes and those of the campus community."[75]

Two days later, the president of the Student Government Association notified the chairman of the board that the Student Senate had "unanimously expressed their opinion that no changes should occur in the bylaws of Assumption College that would seriously threaten the academic integrity and stable operation of the college." He urged the creation of a consultative committee representing all constituencies of the college, including the Assumptionists. "Hopefully," he wrote, "this committee would find a more equitable and reasonable answer to the Assumptionists' perception of problems dealing with their sponsorship." At the trustees' meeting two days later, the chairman of the board asked President Hagan to reconvene the consultation committee and to add several Assumptionists to its membership.[76]

While that process was being organized, the eight-member Assumptionist internal committee responded to the lay consultation committee's report through a letter to the trustees, arguing that the lay people had missed the point. "The central issue," they wrote, "boils down to this: who shares with whom? Is the school an Assumptionist institution in which the laity share broadly with us, or is it a lay institution in which we participate, albeit in a special but undefined way, like all other constituencies of the College?" The statements in the charter and bylaws that the corporation is "under the auspices" of the Assumptionists establish that "we are the ones who do the sharing, not the other way around," the committee said. The members expressed their willingness "to pursue the dialogue through the reconstituted and clearly mandated Consultation Committee requested by the Chairman of the Board."[77]

After consulting with the Assumptionist provincial superior, the president, on May 1, 1981, announced the appointment of a joint consultation committee made up of five Assumptionists and five non-Assumptionists. He charged them with reviewing issues relevant to "effective sponsorship by the Assumptionists at Assumption College" and asked them to report by November 1. After receiving the assistance of two outside consultants, on October 27 the joint committee reported that it had reached a consensus that "the bicameral system should not be adopted by the board." While some members found merit in the idea, all agreed "that the proposal is untimely and could cause more harm than good." The committee added that there should be "further exploration" of structural changes that "might better insure a substantial role for the Assumptionists in major policy decisions." It mentioned two structural possibilities that had been "discussed but not fully evaluated by the committee." One was a suggestion that the provincial superior or his designee might serve ex officio as chairman of the board of trustees; the other possibility raised was that further authority might be vested in the Bylaws and Nominating Committee, whose membership was made up of three lay trustees and three Assumptionists, including the provincial superior. That committee had the exclusive right to propose bylaw changes and to nominate candidates for membership on the board, the Executive Committee, or the Bylaws and Nominating Committee.[78]

The joint consultation committee went out its way to call attention to what it saw as a difference of perception that separated Assumptionists from lay members of the community: "Some in the College community have wondered whether the proposed changes might cause harm to the College while being beneficial to the Assumptionists. The perception among Assumptionists generally is that they are so closely tied to or identified with the College that harm to one would mean harm to the other. The perception among lay collaborators is based on an awareness of the distinction between the two Institutions, Assumption College and the Assumptionists, and of potential for a conflict of interest between the two." Reflecting on its experience, the committee wrote, "We cannot overestimate the major role that misconceptions and contrary perceptions have played in complicating this discussion. The only solution to that problem is regular conversation."[79]

Without further formal consultation with the campus community, the trustees on December 11 amended the bylaws to provide that the North American provincial superior of the Assumptionists would be ex-officio chairman of the Board of Trustees, and they strengthened the Bylaws and Nominating Committee by merging it with — and giving it the title of — the Executive Committee. The new Executive Committee, composed of four Assumptionists and four non-Assumptionists, included, as ex-officio members, the chairman and vice chairman of the board, the president, and the chairman of the board's Finance Committee. In addition to the powers of the preexisting committees, the new Executive Committee was granted the exclusive power to propose resolutions or amendments regarding the mission of the college and the Assumptionist sponsorship, as well as the power to nominate or renominate the president and to propose selection procedures. The revised bylaws also made a change recommended by both President DiPasquale and President Hagan: they excluded employees of the college, other than the president, from membership on the board. President Hagan's announcement of these developments closed with his "hope that these changes will prove satisfactory and that we will now be able to return full-time to the important task of assuring the College's survival during the very uncertain and difficult days ahead."[80]

Part of the agreement reached by the trustees called for a cancellation of residual notes the Assumptionists had been receiving for some years. The notes reflected the annual Assumptionist financial contribution to the college. By June 1981, these documents, which might have given the Assumptionists special claims on college property in the event of liquidation, amounted to $1,072,751. The cancellation was sent from Fr. Loiselle to President Hagan on February 25, 1982.[81]

The Assumptionists also took steps to remedy some of the difficulties caused by the decision in 1972 to locate the long-awaited

Mother Marie-Eugénie Milleret, R.A.

Assumptionist Residence at 50 Old English Road, north of the campus and out of sight. In the mid-1980s, a new residence was constructed to house a new and visible Assumptionist community on the campus. The campus ministry team was to be part of this new community, and all who lived there were to regard the campus as their "apostolic priority." Named Emmanuel House, the new residence was built on land leased from the college, adjacent to the chapel and a short walk from the future Emmanuel d'Alzon Library.[82] From 1986 on, there would be two Assumptionist communities, one at Emmanuel House and one at Old English Road, each with its own superior.

If the definition of "Assumptionists" is used more inclusively, there were actually three Assumptionist communities working at the college by 1986. The second had been founded in 1985, with the arrival — long desired by the Worcester Assumptionists — of members of the Religious of the Assumption. This was the order of nuns founded in Paris in 1839 by Mother Marie-Eugénie de Jésus (Eugénie Milleret). At that time, Mother Marie-Eugénie had already begun what would be a long spiritual collaboration with Fr. Emmanuel d'Alzon, the future founder of the Augustinians of the Assumption. In Worcester, the sisters functioned as students, teachers, and campus ministers, paying special attention to international students. They also organized student visits to their missions in Mexico, Philadelphia, and New York. The sisters were so welcomed by the male Assumptionists that by the end of the college's first century, two of the Assumptionist seats on the college's board of trustees were held by Religious of the Assumption.

Curriculum Reform

Though the core (or general education) curriculum had been revised in 1974–75, further reform had been on the minds of the trustees even before the arrival of President Hagan in 1978. The board's Academic Affairs committee in January of that year adopted a reform schedule. It called for Academic Dean Richard A. Oehling to appoint "a select ad hoc committee of the faculty" and complete a review by September 1978. Following comments by the trustees' committee, the revised curriculum was to be acted upon by the Faculty Senate by March 1979 and by the board in September 1979, with implementation set for September 1980.[83] Few could have imagined that a fully revised set of curriculum requirements would not be in place until September 1989.

That extended process is presented here in some detail for a number of reasons. The primary one is the continuing importance of the general education curriculum shaped during those deliberations. There is also an attempt to capture in context some of the disagreements and misunderstandings, as well as the sense of common purpose, that were part of the atmosphere of the college in the 1980s. The narrative illustrates the faculty's failure to reach internal agreement on curricular reform, the degree of mutual mistrust that existed on some issues between the faculty and the Board of Trustees, and the difficult position the Hagan administration sometimes got into as it tried to move in the rocky middle ground.

Preliminary Inquiries

Following the mandate of the board, Academic Dean Richard A. Oehling, in late January 1978, sent the faculty a written discussion of liberal arts education and a description of six humanities-based approaches to "unified general education programs." He asked each department to discuss these approaches before meeting with him for preliminary discussion about reforming the Assumption curriculum.[84] A few months after President Hagan's arrival, having met with the departments, the dean reported on November 1, 1978, that faculty opinion, while diverse, tended to divide between those wanting to reform the requirements "in fairly modest ways," and those favoring "major reforms." He listed ideas "most commonly expressed" for minor changes in the existing distribution system and presented two outlines for "major reforms." The dean asked that, in the near future, the faculty as a whole, along with the administrators who used to sit in the old Academic Senate, meet and choose whether to undertake minor or major reform. Following that decision, he wrote, that same meeting could elect a committee to "flesh out" the minor or major reform and submit a report to the Faculty Senate, the administration, and the board by March 1, 1979, then four months away.[85]

That request from the dean raised questions among the faculty, including curiosity about "the source of interest in change," questions the senate voted in February 1979 to address to the dean before agreeing to a special meeting.[86] Some weeks later, the senate voted to ask the dean to poll all faculty members concerning their interest in curriculum change. In his December 1979 report to the board, President Hagan declared that the majority of the faculty "do not wish to establish any of the experimental interdisciplinary approaches to general education" proposed by the dean. He said Dean Oehling would instead propose a number of more restricted changes: an increase in philosophy and religious studies requirements from a total of three courses to two courses in each; two semesters of History of Western Civilization for all students; a provision that ethics be one of the two required philosophy courses; and the requirement of a second writing course for students receiving grades below C in the required composition course. By the time the dean's proposals were sent to the faculty on May 2, 1980 , the ethics requirement had been removed, and the grade required to escape a second required writing course had been lowered to C-minus.[87]

On February 6, 1981, the Faculty Senate voted not to accept the addition of one course to the requirement in philosophy and religious studies, nor the two-course history requirement. The senate also turned down the provision for requiring a second course in writing, viewing it as an educationally dubious way to address the problem and preferring to explore what in time became (with Dean Oehling's support) a "writing across the curriculum" program. The senate on March 5, 1982, also acted negatively on a proposal from Dr. Oehling for a single-course requirement in Western Civilization. During the discussion, the senate heard complaints about a "piecemeal" approach to curricular reform and an expression of "dismay that our educational expectations are so often translated into specific course requirements that cannot possibly bear the burden of our mission." At that point, Dr. Charles W. Estus, a sociology professor and director of a grant-funded program to develop interdisciplinary studies, moved that the senate mandate a joint meeting between its curriculum committee and the ten-member interdisciplinary program development committee "to develop a two-year study plan for a college-wide examination of the undergraduate curriculum and instructional programs at the college." The senate approved the motion, thereby launching a faculty-led curriculum review, funded to a considerable extent by the National Endowment for the Humanities. President Hagan reported to the board that the senate and the dean were "in agreement about proceeding with a full-fledged study of the entire general education requirements" and that "the work may constitute the most serious examination of the general educational program since 1973."[88]

Faculty Initiative

At the May 1982 meeting of the board, when the president provided a further report concerning the study process being established by the faculty, concern was expressed about the absence of administrators and trustees from the proceedings, and the president was asked what roles he and the dean would play. President Hagan responded that he would take "a very strong role" and that "the Dean will be the coordinator and will be responsible for the study." The trustees passed a motion that the board and members of its Academic Affairs Committee were to be "involved in the curriculum study, and... be kept advised of its progress."[89]

The assistant dean of academic affairs, Fr. Richard E. Lamoureux, served on what became known as the Curriculum Study Group. The dean of student affairs, Fr. Peter R. Precourt, also joined the group after it had become convinced that many educational issues crossed the boundaries between academics and student life. The Study Group also included six faculty members (two of whom served as co-chairs) and two students. After some months of raising issues and gathering data, the study group "tentatively concluded that there was widespread agreement that Assumption was generally succeeding in its mission as formulated in the college's mission statement and that there was no general demand for changes in the structure of the curriculum." After fall and spring workshops, open to the campus community, the Study Group believed it had heard a "strong consensus on the freshman year as a matter for immediate attention," a project that offered "a good opportunity for first steps in curricular reform." A May 1983 workshop on freshman instruction displayed what the Study Group saw as further consensus on "the importance of attention to academic skills, to developing positive attitudes toward learning, and to developing an appreciation for the interrelatedness of the disciplines," all of which "could provide a stronger foundation from which students could build a successful career in a liberal arts college."[90]

President Hagan reported to the board on May 13 that the study group had decided to focus on "the Freshman year experience." In September, he announced that a further curriculum review would be undertaken "under the leadership of the President and the Dean," and that the work of the existing Curriculum Study Group would be "only the first phase in a complete study of the core curriculum." The president told the board that there were "some serious gaps in the curriculum." According to the minutes, he said this included "the question of whether the students really receive a Catholic, liberal arts education." President Hagan said it was "necessary to have a better sense of what disciplines are necessary to satisfy Catholic liberal arts. The [Curriculum Study Group] talked about everything but the curriculum. Therefore, a committee will be appointed to pursue an in-depth study of the curriculum; it will not be faculty chaired, but some of its members will be faculty." This decision shifted the institutional focus away from the educational experiences of freshmen and back to reforming the structure of the general education requirements. On February 3, 1984, acting on behalf of the board, the Executive Committee approved a resolution, presented by President Hagan, directing the president to create "a College-wide Curriculum Study Committee to review the entire curriculum and to consider possible changes in the core requirements and major areas of concentration."[91]

Trustee and faculty concerns about the curriculum were further stimulated by matters outside the stream of formal curricular reviews. One concern was the success of the business studies program, introduced in 1979, in attracting student majors, a process the president described in March 1982 as "unrelenting."[92] In addition, in 1983 the faculty and administration proposed a major in computer science, an innovation that led to an extensive statement to the trustees by the provincial superior and chairman of the board, Fr. Joseph Loiselle. He warned that the character, mission, and identity of the college could be endangered by "piecemeal program additions and/or subtractions" and asked, "How can we guarantee that Assumption College will continue fulfilling its Catholic Liberal Arts mission in the face of competition, of pessimistic demographic forecasts, and of an increasingly career-oriented student population?" Fr. Loiselle argued that the college needed to be clear "concerning whom we want to recruit and what we want to sell . . . under the banner of Catholic Liberal Arts." He suggested that "the blue-ribbon committee(s) which the President wants to establish" be charged with exploring what sort of person the ideal Assumption graduate would be, the "ideal distribution of major programs, the ideal distribution of the student body in each of those programs," and a core curriculum specifying which courses in the liberal arts must be taken by all students.[93]

Trustee-Mandated Committee

As a result of these steps, by the time the faculty-led Curriculum Study Group published a report of its two-year undertaking in March 1984, planning was well under way for what was sometimes called the "blue-ribbon committee," the Curriculum Study Committee, under the auspices of the board. The study group's central curricular proposal was the establishment of a set of six-credit (two-course) interdisciplinary freshman seminars with a variety of themes, designed to strengthen academic skills, encourage desirable attitudes toward learning, and help students understand the interrelatedness of the disciplines. Each student would be required to choose one seminar in each semester of the freshman year, and the program would not add to, or require changes in, the structure of the general education requirements. In February 1985, the Faculty Senate turned down the proposal after individual members had expressed doubts about several aspects, including interdisciplinary teaching, the willingness of departments to alter their "basic courses," and the potential effects of the seminars on enrollments in some departments. The study group had also recommended a new statement on educational goals and values that was approved by the Faculty Senate and has since that time appeared in the catalog as the "objectives" of the college. The group had also made recommendations in the areas of academic skills, advising and counseling, special events on the campus, campus life, and career counseling, most of which were approved by the senate, referred to relevant offices, or both.[94]

The Faculty Senate at first hesitated to appoint representatives to the new Curriculum Study Committee being set up by the board. Members were concerned about the fact that the faculty's

Curriculum Study Group had not yet reported and wondered whether "the [new] study procedures [would] recognize the traditional role of the faculty in curricular matters." Asked to elaborate on faculty concerns, President Hagan, according to board minutes, "stated that this was not a confrontation situation with the faculty, but it should be approached very carefully. The faculty feel they are principally responsible for the curriculum. They are concerned that the trustees might have a hidden agenda which would give priority to theology and philosophy, and perhaps to English, so that the requirements would detract from the work being done by other departments. Many of the faculty do not see a hierarchy in disciplines which is appropriate to a classical education. The Board should be sensitive to the faculty's feeling that they should make any curriculum changes, but the faculty must also recognize that the Board has the ultimate responsibility for the curriculum and that the President has a key role to play in the entire process."[95]

By May 1984, the board and the Faculty Senate had agreed that there would be nine faculty members on the board's Curriculum Study Committee, rather than the six requested by the board or the fourteen offered by the senate.[96] Also on the committee were three administrators, three trustees, one student, and one alumnus. By the start of the 1984–85 academic year, Fr. Richard Lamoureux, previously assistant academic dean, had become provincial superior, chairman of the board, and chair of the Academic Affairs Committee of the board.[97] In his new position he would have considerable influence on the discussion of the core curriculum.

Fr. Richard E. Lamoureux, A.A.

The Curriculum Study Committee divided itself into three "task forces" to work, respectively, on skills and learning attitudes, philosophy and religious studies, and core requirements in general. In January 1985, the Academic Affairs Committee of the board met without its student and faculty members to review the work of the study committee. It noted that the study committee should "reflect on the rationale governing whatever proposal they make," that President Hagan should "find an occasion to remind the faculty of the Trustees' concern and responsibility for academic policy," and that it would be wise for the Academic Affairs Committee to review in May a draft of the curriculum committee's work and provide it with "a preliminary trustee reaction."[98]

In May, the president reported that the study committee had done a great deal of work, and that he had extended its deadline from May 1 to December 1. The Academic Affairs Committee again reviewed the study committee's work and decided that it "needed a nudge." At President Hagan's suggestion, the committee, and subsequently the board, directed that the president and the dean of academic affairs "draw out the specific [task force] suggestions that seem to have most merit and draft the rationale" to support them, that they send their drafts to the study committee and to the trustee committee — which would send its own suggestions to the study committee. A final report was to be sent to the trustee committee no later than December 1, 1985. This process was intended to lead up to a board vote at its February 1986 meeting.[99]

The preliminary review by the president and the dean appear to have been dropped from the plan, perhaps due to delay at the committee level. In his October 1985 report, President Hagan wrote that the third task force would try to complete its work by early November. "It is hoped and expected," the president wrote, "that the CSC will finish its work... and send its report simultaneously to the faculty and to the Board of Trustees. The faculty will review the recommendations of the CSC and send its recommendation to the Dean of Academic Affairs and the President." At the October 18 board meeting, he said, according to the minutes, that the committee's final report was due December 1 and that "before making his recommendations to the Board, the President will make a judgment on what will be possible with respect to curriculum change. It is important to be sensitive to what the faculty has to say about the curriculum because they are the ones who will have to carry it out and if they are not supportive it will be a failure."[100]

Study Committee Recommendations

The Curriculum Study Committee's final report, dated February 7, 1986, presented a plan that, while continuing to distribute students over a wide range of disciplines, was considerably more prescriptive than the existing core requirements. The committee framed its recommendations in a historical context, claiming that "incremental changes" in programs of study had shifted "the bulk of student course registrations away from traditional majors and courses and into preprofessional programs."[101] The committee believed that while these supposed changes had "strengthened the college and benefitted its students," they had also created a need to "protect the liberal arts character of the college" through revised general education requirements. In addition, based in part on a recent survey in which the faculty had indicated "that incoming students are unprepared to do college-level work," the committee — like the Curriculum Study Group before it — pointed to the need to develop student skills in reading, writing, and critical thinking. The committee thus endorsed the existing English Composition requirement, the planned introduction of a "Writing Emphasis Program" in all majors, the creation of a "learning resource center" to assist students, a critical-thinking component in many courses, and — still under "basic skills" — a foreign language requirement, a mathematics requirement, and a laboratory science requirement for all students.

The committee also recommended requiring two courses in philosophy and two in religious studies. The first course in each was specified (Introduction to Philosophy, The Bible) and the second was to be chosen from a restricted list of intermediate courses. Two courses were recommended in literature, beginning with Introduction to Literature (using "significant literature"), followed by a more advanced course (dealing with "a limited number of major works"). In history the committee recommended a two-course requirement in what came to be called The History of Modern Europe and the United States, and, in fine arts and music, "one course from a short list." In the social sciences, the recommendation was for three courses, one each in three of the eight disciplines.[102]

The new proposal differed from the existing system not only by specifying which courses would fulfill requirements in some disciplines but also in requiring a larger number of courses, although the precise number would vary. For example, whether one or two courses would be required in mathematics or a foreign language would depend upon a student's preparation. Where the 1975 requirements had called for two courses in mathematics or natural science, under the new proposal, a student taking two courses to complete the mathematics requirement would also have to take a laboratory science course. The required total in the humanities rose from four to five, and four courses rather than three would be required in philosophy and religious studies.

Faculty Deliberations, Presidential Recommendations

The Representative Faculty Senate voted to turn the matter over to the full faculty, to be discussed at a series of two-hour meetings on Friday afternoons. The faculty addressed the recommendations of the Curriculum Study Committee sequentially and voted on them in their original or amended forms. It was agreed that these votes were tentative and that they would be followed by a final vote on the full proposal, as amended. In general, the faculty supported the Curriculum Study Committee's recommendations, although some votes on requirements were closely divided and some changes were voted. The eighth faculty meeting, on May 22, culminated in a final vote by which the overall proposal was rejected on a vote of nineteen in favor, twenty-one opposed, and four abstentions. Sensitive to the trustees' interest in bringing the long process to an end, the faculty before dispersing for the summer voted to ask the board not to take action before its February 1987 meeting. During the intervening semester, the resolution said, the faculty would "produce a curriculum that is satisfactory to the faculty" or would "detail the substantive issues which divide the faculty."[103]

On June 27, President Hagan sent the trustees his promised report on curriculum revision. He observed that objections among faculty members to added requirements tended to be based both on educational principle and on concern that too many requirements might reduce the college's competitiveness. The president agreed on most points with the Curriculum Study Committee recommendations, though he argued for three rather than two required courses in theology. On September 16, Fr. Richard Lamoureux submitted to the board the curricular recommendations of the trustees' Academic Affairs Committee, which he chaired. Although generally supportive of the Curriculum Study Committee's recommendations, they called for a total of five rather than four courses in philosophy and religious studies. In a closing statement on "general principles," Fr. Lamoureux stated that "trustees have a special concern to assure that a Catholic Liberal Arts curriculum exists in fact at Assumption College, and not just in theory." He listed, as the disciplines that must hold "a central place in the curriculum," theology, philosophy, history, literature, language, fine arts, natural and social sciences, and mathematics. (Of the courses regularly offered by the college, those not included were in business studies, education, and computer science.) He also stated that courses in the core curriculum should "emphasize the major writers/thinkers/artists of the Western tradition and their major works."[104] The trustee vote on the new curriculum was scheduled for October 17, 1986.

On the preceding afternoon, a special meeting of the faculty was called to discuss the apparent intention of the trustees not to honor the faculty's request for a one-semester delay, as well as the action of the trustees' Academic Affairs Committee in proposing alterations in (tentative) faculty recommendations. The faculty members present voted to declare that curriculum decisions "lie within the exclusive province of the faculty subject to approval but not revision by the Board of Trustees." They also declared that the president's submission of "an independent curricular proposal" and the trustees' "overriding the curricular proposal of the faculty" indicated "bad faith or lack of confidence in the faculty."[105]

Trustee Decisions

At the trustees' meeting the following day there was strong agreement on the importance of reasserting the trustees' authority over the curriculum and of giving solid support to the president. The board unanimously endorsed a statement that the board supported the president "in the leadership role that he has played as a result of our requesting him to play that role, and we continue to support him in whatever decisions he makes and actions he needs to take in the future to implement the curriculum." It was also decided that the board should address itself to the president, and not the faculty, informing him of its approval of the curriculum as submitted by its own committee.[106] The October 21 letter to the president, signed by the chairman of the board, included the observation that the board had, two years earlier, asked the president not only to oversee a curriculum review but also to present his own recommendations to the board.[107]

As early as October 27, a number of academic departments began entering into the record of the Faculty Senate statements disassociating themselves from all or part of the October 16 faculty vote, especially the reference to "bad faith," which President Hagan had described as "devastating" and "perhaps the most serious allegation one can make against another human being."[108] Despite these declarations, the senate did not take steps to change or withdraw the action taken by the full faculty. However, in January 1990, when President Hagan had been offered the presidency of Rider College in Lawrenceville, N.J., the senate did vote overwhelmingly to express its gratitude and appreciation to him, as well as its hope that he would remain at Assumption College.[109]

Following the trustees' adoption of new general education requirements, the Faculty Senate established a standing committee on general education to "evaluate and make recommendations concerning the curriculum" and also to serve as an advisory committee to the dean of academic affairs regarding administrative issues arising during implementation of the new program. The goal of the trustees was to have decisions on implementing the curriculum taken by the end of the 1986–87 academic year so that — with allowance for publishing the requirements and recruiting freshmen under the new program — the new curriculum might go into effect in September 1988. The committee on general education set up three subcommittees. One was to look into the competency levels to be established in such areas as foreign languages and mathematics (as well as means of exempting students already at those levels from having to take required courses); a second was to examine the financial implications, including the hiring of new faculty members and the availability of classrooms and laboratories; the third was to study the likely effects of the additional requirements on transfer students and students enrolled in especially time-consuming programs, such as the concentration in education, study abroad, double majors, internships, and the business option in economics. As it met during the spring semester of 1987, the committee discovered that it would not be possible to carry out its studies, hold discussions with the faculty, and make recommendations by the end of the academic year. At the urging of the administration, the trustees on May 22 reluctantly agreed to allow the implementation date to be set back to September 1989.[110]

Finally, on December 1, 1987, the committee on general education recommended to the faculty senate several proposed revisions in the trustee-mandated curriculum. To add flexibility, the committee recommended that one of the five required courses in philosophy and religious studies, one of the two in history, and one of the two in literature be turned into options students might choose among to complete a three-course "depth requirement." It also recommended that students be required to take courses (or establish competency) in two of the three areas of mathematics, foreign languages, and natural sciences rather than in all three. The senate approved those recommendations on December 7. After a meeting between the committee on general education and the Academic Affairs Committee of the board, the board on February 26, 1988, approved the faculty revisions. The new general education curriculum, originally mandated for September 1980, was implemented in September 1989. With some modifications, it continues to serve as the structure of core requirements for all students.

Surviving the Demographic Curve

At the close the 1989–90 academic year, Assumption's supporters could once again celebrate evidence of the college's growth and prosperity. The student body (1,837) was larger than ever before, and the operating budget had generated excess revenues of $1,741,810. The number of applicants for the next freshman class had decreased by 7 percent, but that was, nevertheless, the second largest applicant pool in the college's history. Some concern was raised early in the summer when attendance at orientation sessions was below expectation, and the evidence of a problem hardened when fifty-five students fewer than expected arrived for classes. In retrospect, it was observed that, in addition to the fact that the demographic dip in the number of people between eighteen and twenty-two still had around five years to run, the local, state, and national economies were in a deeper recession than anticipated, and Assumption and competing institutions had all been involved in a scramble to fill up their student ranks. For the first time since the early years of the Hagan presidency, urgent belt-tightening was required across the campus departments, and the budget for the 1991–92 year anticipated a further reduction in the number of students.[111]

For the next few years, policy makers and administrators devoted a great deal of attention to means of recruiting new students and keeping them at the college once they had arrived. A "learning center" where students could receive help with reading, writing, and study skills, endorsed by all parties during the curriculum debates, was soon functioning as the Academic Support Center in the Emmanuel d'Alzon Library. New forms of academic advising and new programs to assist the most "at-risk" freshmen were also established. The trustees and the administration also took steps to address the ever-increasing need for financial aid among prospective students.

In 1990, Fr. Richard E. Lamoureux completed his second three-year term as provincial superior and chairman of the board and became provost of the college, succeeding Dr. Richard A. Oehling, who had served as chief academic officer since 1973. Rowing somewhat against the financial tide, Fr. Lamoureux successfully advocated a phased-in reduction in the teaching loads of full-time faculty (from eight courses per year to seven) to allow more time for scholarly development as well as service to students. Fr. Lamoureux also developed a program of full-year sabbatical leaves, to be awarded to faculty members on a competitive basis. (Nonwinners continued to receive sabbatical leave for one semester at full salary or one year at half salary.) In 1995, serving as acting president while President Hagan was on leave, the provost also participated in the development of new merit scholarships, intended to improve the academic quality of the student body.[112]

That same year, Treasurer Francis P. Gurley announced that the college was out of the "demographic doldrums" of the years between 1979 and 1994, and that the supply of prospective students had begun to increase. Although 1994–95 had been "a very difficult year," the college could look forward to a better recruiting environment.[113] During the 1996–97 academic year, President

Dr. Charles L. Flynn, Jr.

Hagan and a new provost, Dr. Charles L. Flynn, Jr., hired a consulting firm to help guide the transition in the admissions operation from the leadership of Thomas Dunn, who was retiring as vice president for enrollment management after twenty-nine years of service, twenty-seven of them as head of the admissions department. The consultants' focus was on a "financial aid leveraging system" to help direct financial assistance toward the most academically qualified students, as well as on other "new recruiting techniques." In May 1997, the president could announce that, although the number of applicants has been about the same as in the previous year, around 575 freshmen could be expected to enroll in the fall, the largest new class ever.[114] In fact, 597 freshmen enrolled in September.

"I genuinely believe that we are turning a corner," the president told the trustees in May. "I am beginning to suspect that Assumption is about to become a 'hot' school." He suggested that the student body of 1,650 students be allowed to grow to 1,850, within the next two years and that a firm be hired to develop a master plan for the campus. He also requested and received approval for the immediate construction of a new residence (the Living/Learning Center). "This is an exciting time in the College's history," President Hagan concluded. "I am confident that we can address the short-term challenges [overcrowding] posed by the success of this year's recruitment efforts. At the same time, I am prepared to move forward immediately to develop a Master Plan to guide the growth of the College into the new millennium."[115]

On the first day of classes the following September, members of the college community found in their mailboxes a message from the president that began, "As most of you know, I had planned to remain as President of Assumption College until July 31, 2000. A decision has been made to move up that date to July 31, 1998, when I shall have completed twenty years as President of the College." The memorandum did not indicate who had made the decision or why. When asked by a local newspaper reporter, the president said he had reconsidered his plan to remain at Assumption until he reached the age of 65. "With two more years to go," he told the reporter, "I began to get the feeling I was a lame duck. It was a feeling I was getting from some faculty members and trustees." According to a report in the diocesan newspaper, *The Catholic Free Press,* President Hagan "said... that by advancing the date of his retirement, he would be able to pursue other opportunities in higher education, business, and public service."[116]

Fr. John L. Franck, A.A.

No further explanation emerged officially from President Hagan or from the Board of Trustees during the final year of the president's tenure — which was shortened to one semester. In December 1997, the president, in a message to the community, wrote that he had decided, with the permission of the board of trustees, to begin a promised "year-long terminal sabbatical" in January 1998 because he "would like to have at least six months for rest, relaxation, and renewal prior to undertaking any new set of responsibilities." On the same day, the community received a message from Fr. John L. Franck, provincial superior and chairman of the board, announcing that the provost, Dr. Charles Flynn, and the executive vice president (and treasurer), Francis Gurley, would "carry out jointly the responsibilities of the presidency beginning January 1, 1998" and serve until a successor to President Hagan had been named.[117]

At the invitation of the senior class, President Hagan returned to the campus as president emeritus in May to deliver the commencement address. On that occasion, both he and Patrice O'Malley Hagan were awarded honorary doctorates in recognition of their years of service to the college and to the outside community, in which both Hagans had been very active during their two decades in Worcester. The spring 1998 issue of *Assumption College Magazine,* was largely dedicated to "Reflections on the Hagan Legacy, 1978-1998."

Notes

ABBREVIATIONS

AC — Archives of Assumption College, Worcester, MA

AG — Archives of the Assumptionist General House, Rome

AP — Archives of the Assumptionist Provincial House, Brighton. MA

PA — Archives of the Assumptionist Province of Paris

SP — Archives of the Diocese of Springfield, MA

1. Trustee Minutes, June 2, 1972, AC.
2. Trustee Minutes, Jan. 5, 1973, AC.
3. Trustee Minutes, Sept. 29, 1972, AC.
4. Trustee Minutes, Sept. 29, 1972; Jan. 5, 1973. AC.
5. Trustee Minutes, Sept. 29, 1972; Jan. 5, 1973; Sept. 29, 1972, AC.
6. Trustee Minutes, Sept. 29, 1972; Jan. 5, 1973; Sept. 29, 1972; May 26, 1973, AC.
7. *Assumption News* 35, 5 (April 1974): 2; Trustee Minutes, Jan. 17, 1975, AC.
8. Trustee Minutes, Feb. 25, 1977; Sept. 30, 1977, AC.
9. Trustee Minutes, March 23, 1973, AC.
10. Trustee Minutes, Jan. 11, 1974, AC.
11. DiPasquale, Report to the Board of Trustees, Sept. 27, 1974, filed with Trustee Minutes, AC.
12. Trustee Minutes, Sept. 27, 1974, Jan. 17, 1975, AC.
13. Loiselle, "Report of the Student Life Committee," Dec. 20, 1974, filed with Trustee Minutes, Jan. 17, 1975; Trustee Minutes, Jan. 17, 1975, AC.
14. DiPasquale, Report to the Board of Trustees, Sept. 19, 1975, filed with Trustee Minutes; Minutes, Student Life Committee of the Board of Trustees, Sept. 19, 1975, filed with Trustee Minutes; DiPasquale, Report to the Board of Trustees, May 14, 1976, filed with Trustee Minutes, AC.
15. Trustee Minutes, Sept. 27, 1974, AC.
16. DiPasquale, Report to the Board of Trustees, May 14, 1976, filed with Trustee Minutes, AC.
17. Trustee Minutes, Sept. 28, 1973; Sept. 27, 1974; Engel, Report to the Executive Committee, Oct. 28, 1975, filed with Trustee Minutes, Dec. 5, 1975, AC.
18. Trustee Minutes, March 21, 1975; Dec. 5, 1975; Feb. 27, 1976, AC.

19. Trustee Minutes, Dec. 10, 1976; DiPasquale, Report to the Board of Trustees, May 20, 1977, filed with Trustee Minutes; Wilfrid J. Dufault, Report to the Board of Trustees, Sept. 30, 1977, filed with Trustee Minutes, AC.

20. Trustee Minutes, Jan. 5, 1973; May 26, 1973, AC.

21. Robert J. Fortin, Report of the Student Life Committee, April 29, 1975, filed with Trustee Minutes, May 16, 1975, AC.

22. Trustee Minutes, Dec. 5, 1975, AC.

23. Report of the Academic Affairs Committee, May 21, 1974, filed with Trustee Minutes, May 25, 1974, AC.

24. Trustee Minutes, March 21, 1975, AC.

25. Trustee Minutes, May 16, 1975, AC.

26. Trustee Minutes, March 21, 1975, AC.

27. C. Brusard to DiPasquale, "Senate response to curriculum recommendations by the Board of Trustees," April 23, 1975, filed with Trustee Minutes, May 16, 1975; Trustee Minutes, May 16, 1975, AC.

28. DiPasquale, Report on the Trustee Seminar Weekend from the President, Sept. 1, 1976, filed with Trustee Minutes, Sept. 17, 1976; Trustee Minutes, Sept. 17, 1976; DiPasquale, Report to the Board of Trustees, May 20, 1977, filed with Trustee Minutes, AC.

29. Trustee Minutes, Feb. 25, 1977, AC.

30. Draft Presidential Profile, April 28, 1977, filed with Trustee Minutes, May 20, 1977, AC.

31. DiPasquale, Report to the Board of Trustees, May 20, 1977, filed with Trustee Minutes, AC.

32. DiPasquale, Report to the Board of Trustees, May 20, 1977, filed with Trustee Minutes; Trustee Minutes, May 20, 1977, AC.

33. Dufault, Report to the Board of Trustees, Sept. 30, 1977, filed with Trustee Minutes, AC.

34. Dufault, Report to the Board of Trustees, Sept. 30, 1977, filed with Trustee Minutes, AC.

35. Trustee Minutes, Dec. 4, 1977, AC.

36. "Address by President Joseph H. Hagan to the College Community at the opening Convocation and Mass of the Holy Spirit, September 14, 1978," filed with Trustee Minutes, Oct. 6, 1978, AC.

37. Trustee Minutes, Oct. 6, 1978, Dec. 8, 1978, AC.

38. Hagan, Report to the Board of Trustees, March 16, 1978, filed with Trustee Minutes; Trustee Minutes, May 19, 1979; Hagan, Report, Oct. 5, 1979, filed with Trustee Minutes; Trustee Minutes, Dec. 7, 1979; Hagan, Report, March

14, 1980, filed with Trustee Minutes, AC.

39. Trustee Minutes, May 15, 1981; Hagan, Report, Oct. 2, 1981, filed with Trustee Minutes; Trustee Minutes, March 18, 1983, AC.
40. Hagan, Report, March 18, 1983, filed with Trustee Minutes; Hagan, Report, Feb. 8, 1985, filed with Trustee Minutes; Trustee Minutes, Feb. 8, 1985, AC.
41. Trustee Minutes, Feb. 24, 1989, AC.
42. Hagan, Report, Feb. 23, 1990, filed with Trustee Minutes, AC.
43. Hagan, Report, Oct. 18, 1991, filed with Trustee Minutes; *Assumption Quarterly* 15 (Winter 1992/93): 2, AC
44. Trustee Minutes, May 17, 1996, AC.
45. *Worcester Telegram & Gazette*, Nov. 22, 1993, B6.
46. Trustee Minutes, March 17, 1978; Hagan, Report, Oct. 6, 1978, filed with Trustee Minutes; Trustee Minutes, Oct. 18, 1985; Financial Update, February 1991, filed with Trustee Minutes, Feb. 22, 1991; Hagan, Report, Oct. 6, 1978, filed with Trustee Minutes; *Assumption Quarterly* (Oct. 1983): 3–4.
47. Trustee Minutes, Oct. 6, 1978, AC.
48. Trustee Minutes, March 16, 1979, AC.
49. Trustee Minutes, Oct. 12, 1984, AC.
50. Trustee Minutes, Oct. 18, 1985, AC.
51. Trustee Minutes, Oct. 17, 1986, AC.
52. Hagan, Report to the Board of Trustees, Feb. 26, 1988, filed with Trustee Minutes, AC.
53. Hagan, Report to the Board of Trustees, Oct. 13, 1989, filed with Trustee Minutes, AC.
54. Hagan, Address, Sept. 14, 1978, filed with Trustee Minutes, Oct. 6, 1978; Trustee Minutes, Oct. 6, 1978; Hagan, Report, Oct. 5, 1979, filed with Trustee Minutes; Bylaws and Nominating Committee to Board of Trustees, July 23, 1980, filed with Trustee Minutes, Oct. 3, 1980, AC.
55. Hagan, Report, Oct. 6, 1978, filed with Trustee Minutes; Trustee Minutes, March 16, 1979; March 14, 1980, AC.
56. Minutes, Buildings and Grounds Committee, filed with Trustee Minutes, March 14, 1980, AC.
57. Trustee Minutes, March 20, 1981, AC.
58. Trustee Minutes, Oct. 2, 1981; Hagan, Report to the Board of Trustees, Sept. 30, 1983, filed with Trustee Minutes, AC.
59. Trustee Minutes, Oct. 12, 1984; Feb. 8, 1985; Hagan, Report, May 17, 1985, filed with Trustee Minutes, AC.

60. Trustee Minutes, Oct. 17, 1986. AC.
61. Trustee Minutes, Oct. 16, 1987; Feb. 26, 1988, AC.
62. Trustee Minutes, May 20, 1988; Feb, 24, 1989, AC.
63. Trustee Minutes, May 20, 1988; Feb. 24, 1989; Oct. 13, 1989, AC.
64. Agreement, Assumption College and the Augustinians of the Assumption, Inc., Hagan Presidential Papers, AC.
65. Trustee Minutes, May 19, 1979, AC.
66. Trustee Minutes, March 16, 1979, AC.
67. Trustee Minutes, Oct. 5, 1979, AC.
68. Trustee Minutes, Oct. 5, 1979, AC.
69. Trustee Minutes, Oct. 5, 1979; Excerpt from Minutes, Council of the Province, Oct. 25, 1979, [and] April 2, 1980, Hagan Presidential Papers, AC.
70. Senate Minutes, Nov. 2, 1979; Feb. 20, 1980; Mar. 28, 1980; April 18, 1980; Sept. 12, 1980, AC.
71. Robert J. Fortin, "The Assumptionists and Assumption College," April 2, 1980, Hagan Presidential Papers, AC.
72. Fortin, "Talk to the Faculty Senate, Oct. 17, 1980," Hagan Presidential Papers, AC.
73. Fortin, "Revision of the Bylaws," Dec. 10, 1980, Hagan Presidential Papers; Fortin to Planning Committee and Consultation Committee on Restructuring, Jan. 23, 1981, Hagan Presidential Papers, AC.
74. Hagan to Faculty, Administration, Student Senate, Jan. 26, 1981, Hagan Presidential Papers; Covering Statement of the Lay Consultation Committee Regarding the Proposed Changes to the Bylaws of the College, March 16, 1981, Hagan Presidential Papers, AC.
75. Covering Statement of the Lay Consultation Committee, March 16, 1981, Hagan Presidential Papers, AC.
76. Jeffrey P. Hayden to Richard Testa, March 18, 1981, Hagan Presidential Papers; Trustee Minutes, March 20, 1981, AC.
77. Internal committee to members of the Board of Trustees, April 14, 1981, Hagan Presidential Papers, AC.
78. Hagan to Assumption College Community, May 1, 1981; Hagan to Joint Consultation Committee, May 1, 1981, Hagan Presidential Papers; Report of the Joint Consultation Committee, Oct. 27, 1981, Hagan Presidential Papers, AC. The author of this history was co-chair of the Joint Consultation Committee.
79. Report of the Joint Consultation Committee, Hagan Presidential Papers, AC.

80. Hagan to Faculty, Administrators, and S.G.A. Officers, Dec. 14, 1981, Hagan Presidential Papers, AC.
81. Loiselle to Hagan, Feb. 25, 1982.
82. Henry Moquin and Richard Richards, *Assumptionists in the United States,* 23.
83. Minutes of Meeting of Academic Affairs Committee, Jan. 25, 1978, filed with Trustee Minutes, March 17, 1978, AC.
84. Oehling to Department Chairs, Jan. 20, 1978, AC.
85. Oehling to Department Chairs, Jan. 20, 1978; Oehling to All Faculty, Nov. 1, 1978, AC.
86. Senate Minutes, Feb. 1, 1979, AC.
87. Senate Minutes, March 27, 1979; Hagan, Report, Dec. 7, 1979, filed with Trustee Minutes; Oehling to All Faculty, Representative Senate, Senior Administrators, S.G.A., May 2, 1980, filed with Trustee Minutes, May 16, 1980, AC.
88. Senate Minutes, Jan. 30, Feb. 6, 1981; March 5, March 26, 1982; Hagan, Report, March 26, 1982, AC.
89. Trustee Minutes, May 21, 1982, AC.
90. Report and Recommendations of the Curriculum Study Group, March 1984, 7, AC. The author of this history served as co-chair of the Curriculum Study Group.
91. Hagan, Report, May 13, 1983, filed with Trustee Minutes; Hagan, Report, Sept. 13, 1983, filed with Trustee Minutes; Trustee Minutes, Sept. 30, 1983; Executive Committee Report, March 9, 1984, filed with Trustee Minutes, March 23, 1984, AC.
92. Hagan, Report, March 26, 1982, filed with Trustee Minutes, AC.
93. Hagan, Report, March 26, 1982, filed with Trustee Minutes; Trustee Minutes, Dec. 9, 1983, AC.
94. Report and Recommendations of the Curriculum Study Group, March 1984, 27; Senate Minutes, Feb. 4, Nov. 16, 1985, AC.
95. Senate Minutes, March 2, 1984; Trustee Minutes, March 23, 1984, AC.
96. Trustee Minutes, March 23, 1984; Hagan, Report, May 11, 1984, filed with Trustee Minutes, AC.
97. Trustee Minutes, Oct. 12, 1984, AC.
98. Hagan, Report, Feb. 8, 1985, filed with Trustee Minutes; Minutes, Academic Affairs Committee, Jan. 15, 1985, filed with Trustee Minutes, Feb. 8, 1986, AC.
99. Trustee Minutes, May 17, 1985, AC.

100. Hagan, Report, Oct. 18, 1985, filed with Trustee Minutes; Trustee Minutes, Oct. 18, 1985, AC.

101. In fact, during the spring 1986 semester, the student "head count" in courses in business and in education amounted to 13 percent of the total course enrollments. If Social and Rehabilitation Services — ordinarily placed in the category of the social sciences — were instead included among "preprofessional programs," the head count in that category would have risen to 19 percent of the total.

102. Curriculum Study Committee, Final Report, Feb. 7, 1986, AC.

103. Senate Minutes, Feb. 10, 1986; Lance W. Carluccio to Hagan and Lamoureux, June 5, 1986, AC.

104. Hagan, Report, June 27, 1986, filed with Trustee Minutes, Oct. 17, 1986; Lamoureux to Board of Trustees, Sept. 16, 1986, filed with Trustee Minutes, Oct. 17, 1986, AC.

105. "Motion from the Faculty, Oct. 16, 1986," Hagan Presidential Papers, AC.

106. Trustee Minutes, Oct. 17, 1986, AC.

107. Trustee Minutes, Oct. 17, 1986; Lamoureux to Hagan, Oct. 21, 1986, Hagan Presidential Papers, AC.

108. Senate Minutes, Oct. 26, 1986; Hagan, "Remarks at President's Fall Faculty Meeting, Oct. 24, 1986," Hagan Presidential Papers, AC.

109. Senate Minutes, Oct. 26, 1986; Hagan, "Remarks at President's Fall Faculty Meeting, Oct. 24, 1986," Hagan Presidential Papers; Senate Minutes, Jan. 29, 1990, AC.

110. Senate Minutes, Dec. 8, 1986; Trustee Minutes, May 22, 1987, AC. The author of this history served on the committee.

111. [Gurley], Financial Update, Feb. 1991, attached to Finance Committee report, Feb. 13, 1991, filed with Trustee Minutes, Feb. 22, 1991, AC.

112. Trustee Minutes, Oct. 13, 1995, AC.

113. Trustee Minutes, Oct. 13, 1995, AC.

114. Trustee Minutes, Oct. 13, 1995; Trustee Minutes, May 16, 1997, AC.

115. Trustee Minutes, May 16, 1997, AC.

116. Hagan to Assumption College Community, Sept. 2, 1997, AC; *Telegram & Gazette,* Sept. 3, 1997, 1,16; *The Catholic Free Press,* Sept. 5, 1997, 1.

117. Hagan to Assumption College Community, Dec. 10, 1997; Franck to Assumption College Community, Dec. 10, 1997, AC.

Epilogue
Past into Present

In the course of daily life, the present slips inexorably into the past. In a historical account of a living institution, however, time travels in the other direction. The past must eventually be permitted to flow into the present.

The current era in the life of Assumption College may usefully be dated from April 17, 1998, when Worcester's daily newspaper announced in a front-page story, "Assumption Trustees Offer Plough Top Post." Thomas R. Plough, president of North Dakota State University, had been offered the Assumption presidency on the previous day. According to the news account, the Assumption trustees had expected a reply by 10:30 the previous night but had heard nothing. In Fargo, meanwhile, Dr. Plough was hearing the voices of nearly 400 students and faculty members who had pitched tents and set up cooking grilles outside his campus home in an effort to persuade him to remain. The next day, around 1,000 people staged a rally in the student union. The report of a sign proclaiming, "You Can't Have an Agricultural College without a Plough" helped Worcester readers learn how to pronounce the name of the fifty-seven-year-old man who, a couple of days later, announced that he would be coming to Salisbury Street as the fifteenth president of Assumption College.[1]

Dr. Plough had served North Dakota State as its president for three years. A sociologist by academic training, he had previously served as provost and executive vice president at the Rochester Institute of Technology. He and his family moved into the president's house at Assumption in the summer of 1998.

The new president soon began to coin a repertoire of expressions to communicate his vision of the college, its mission, and its future, a collection of words and phrases from which he drew repeatedly, all the while adding new elements as his ideas evolved. During his first year, he said frequently that, with proper support and leadership, "Assumption College will move from a well-regarded little r, little i, regional institution in New England to a highly-regarded big R, big I Regional Institution in the Northeast." He adopted a slogan already appearing in the publications of the admissions office — Learn. Achieve. Contribute. — often pointing to it as "a theme which emphasizes behaviors and values consistent with our motto, 'Until Christ be formed in you.'" He frequently characterized Assumption as "a high-impact, value-driven, Catholic residential, liberal arts and sciences college" and as a "learning community."

Early in 2000, the Plough administration announced plans for a "strategic expansion of the campus to strengthen our competitive position well into the future." The administration anticipated "a slow but steady increase in the number of graduating high school seniors in most of its major market areas over the next eight to ten years." The strategic plan was designed to take advantage of that increase to further enhance the academic caliber of incoming students. By increasing the size of the undergraduate student body from 1,850 students to 2,150 and providing more attractive facilities, the college hoped to recruit increasingly talented students. The first phase, under way by the spring of 2000, included two new residence halls behind the Living/Learning Center (North Hall and South Hall), seven classrooms and art stu-

Dr. Thomas R. Plough and Mrs. Monty Plough

dios attached to the Laska Gymnasium, renovation of the Bishop Wright Wing of Taylor Dining Hall to accommodate the Office of Institutional Advancement, the relocation of Student Health Services to a redesigned Armanet Hall, and a renovation of the snack-bar portion of the Hagan Campus Center — known then as the Junction, later as Charlie's — including the addition of a new patio and glass entrance. The second phase would involve a third new dormitory (West Hall) and the construction of a large new science center on Chapel Road.[2]

As the first phase of construction was being implemented and the campus master plan developed, phase two of "strategic positioning" took on new elements, including two parking decks and the construction of a new Information Technology Center attached to Fuller Hall. Most of the projects were financed by borrowing, with an eye toward a capital fund drive to be conducted in connection with the college's 100th anniversary in 2004. In January 2001, the capital campaign was already in its "quiet phase."

The formal public announcement of the Centennial Campaign took place in Worcester's stately Mechanics Hall on October 19, 2002. The campaign's goal was set at $30 million, to be raised by May 31, 2006. The goal includes $15 million for capital projects, $10 million for endowment funds, and $5 million in contributions to the annual fund.

It will be the task of future historians to place the events of these years in historical perspective. This brief excursion into current events can close with a recent formulation of President Plough's vision of Assumption and his vision for Assumption. It was published in *Assumption Magazine,* in January 2003, a century after the Assumptionists were negotiating to establish an industrial, commercial, and agricultural school — and an alumnate — in Granby, Massachusetts.

> *This fall, I titled my opening address "Sustaining Engagement and Civility." I wanted to celebrate the hallmarks of this campus, which are friendliness, civility, collaboration, colleagueship, a sense of community, and a commitment to mission. These hallmarks undergird our mission as a coed, residential, comprehensive, liberal arts and professional studies College, founded by the Assumptionists in the Catholic intellectual tradition. The College was and is designed to graduate individuals whose values and behavior are true to our "Learn. Achieve. Contribute." theme, and whose educational outcomes of competence, character and compassion are outward signs of our internalized motto of "Until Christ be formed in you." We value a balance of studies in both the traditional liberal arts and sciences disciplines and complementary professional fields of study, wherever student demand, retention, and post-graduate success make them fiscally viable and intellectually credible. At the same time, in a society that is narrowly career oriented and focused on vocationally relevant programs and skills, it is important to maintain a liberal arts foundation in all that we do and to demand the same emphasis on achieving broad perspectives as on developing specific competencies.*

As a college established in the Catholic tradition, we do have high expectations for student behavior, both in and out of the classroom. Just as assignments for courses are expected to be completed, campus requirements to act civilly, obey the law, and respect property and persons are expected to be honored. Assumption College has performance standards for civility and respect for others, which are just as important as academic achievement. Our student body handles both its in-class and on-campus responsibilities very well, and is a group of young men and women any campus would welcome and celebrate. Perhaps the event this fall which best captured the ethos of the campus and the basic goodness of our students was the candlelight march and remembrance garden dedication to those lost in the terrorist attack of 9-11-01. It was an awesome display of caring and prayer.

This fall, I had the honor of announcing a $30 million Centennial Campaign, which will conclude in 2006 and span our hundred-year birthday party celebration in 2003–2004. Our centennial celebration events and our Centennial Campaign fund-raising activities will offer many additional opportunities to tell others about the renaissance under way at Assumption College. We are very good at taking good students and giving them a great education. We focus on the question of how to live a life in order to find satisfaction out of making a living. Everywhere I go, people tell me that our alumni are good neighbors, involved citizens in their community, faith-based individuals, and just nice people to be around. This is a great story to tell as we strategically position Assumption College from a well-regarded institution in New England to a highly regarded institution in the Northeast.

This narrative began with the unlikely convergence in Worcester, at the dawn of the twentieth century, of French Assumptionists and French-Canadian immigrant families eager to educate their sons. Although neither the French Assumptionists nor the Canadian families are part of the scene a century later, the institution born in Greendale as the Alumnate of Our Lady of Consolation lives on at 500 Salisbury Steet as Assumption College. In 1904, during the first weeks of instruction at the alumnate, there was only one professor, an Assumptionist priest named Fr. Donat Cordier. The undergraduate catalog for 2002–2003 listed five Assumptionist priests teaching at the college, only one of them teaching full-time. He, too, was named Donat: Fr. Donat R. Lamothe, professor of music and archivist of the college. The twenty-first-century Fr. Donat, however, was not bereft of full-time faculty colleagues. There were 150 listed in the catalog, fifty-six of them women.[3]

Fr. Donat Lamothe lives in a society greatly changed from the one into which Fr. Donat Cordier and the other French Assumptionists migrated. This study has explored how the Assumptionists and their co-workers navigated the turbulent seas of the twentieth century and, not only kept their college alive, but brought it in good health to its centennial anniversary.

One theme that emerges from the records is the difficult but crucial role of the American-born Assumptionists in the college's transition from its remarkable past as a small, French-designed, ethnic institution, sometimes insular in outlook, dispensing a rigorous classical education, and staffed mostly by religious; to its also remarkable present as an American-made "comprehensive, liberal arts and professional studies college" — to use President Plough's description — faithfully proclaiming its Catholic character, staffed almost entirely by lay people, and open to the currents of modern scholarship flowing through the general education program, more than thirty major programs in the undergraduate college, and many others in the continuing education and graduate programs. With important help from lay collaborators, the American

Fr. Donat R. Lamothe. A.A.

Assumptionists, to preserve the college's work, had to face and surmount the trying and sometimes painful challenges associated with delay in the transition from French to Franco-American leadership; the shedding of the exclusively French ethnic identity; the trauma of the tornado; the spartan conditions of pioneer life on Salisbury Street; the intellectual, political, and religious turmoil of the 1960s; the professionalization, laicization, and diversification of the faculty; the changeover to a lay-clerical partnership on the Board of Trustees; the loss of Assumption Prep; the financial crisis of 1970–71; and the advent of lay presidents.

Of course, earlier generations of Assumptionists had also met and surmounted formidable challenges: the Granby fiasco; reluctant and shaky early leadership; mistrust from the bishop of Springfield; the World War I mobilizations; the internal strife of the postwar years; the fire of 1923; the dangers posed by interethnic conflict in the surrounding society; the threats of debt and economic depression; and the virtual disappearance of the college-level student body during World War II.

When Fr. Armand Desautels was working tirelessly to insure the survival of his beloved college after the tornado, he used to take comfort in saying that the college's welfare was being placed in the hands of Emmanuel d'Alzon, founder of the Augustinians of the Assumption. Many of Fr. Desautels' co-workers have also expressed confidence in Fr. d'Alzon's capacity to intercede before God in the interest of his sons (and eventually his daughters) and their college in Worcester. The Assumptionists' faith in their founder is such that the Congregation has long sought to advance the cause of his canonization. Missing so far has been evidence of one or more miracles that might be ascribed to his intercession. Readers of this history might have some suggestions along those lines.

Notes

1. *Worcester Telegram & Gazette,* April 17, 1998, 1A; April 21, 1998, 1A.
2. *Assumption* (Spring 2000): 8–9.
3. *Assumption College Undergraduate Catalog,* 2002–2003, 177–83.

Index